Practical Applications
Plant Molecular Biology

Practical Applications of Plant Molecular Biology

R. J. Henry

Professor of Plant Conservation Molecular Genetics
at Southern Cross University, Australia

CHAPMAN & HALL

London · Weinheim · New York · Tokyo · Melbourne · Madras

Published by Chapman & Hall, 2–6 Boundary Row, London SE1 8HN

Chapman & Hall, 2–6 Boundary Row, London SE1 8HN, UK

Chapman & Hall GmbH, Pappelallee 3, 69469 Weinheim, Germany

Chapman & Hall USA, 115 Fifth Avenue, New York, NY 10003, USA

Chapman & Hall Japan, ITP-Japan, Kyowa Building, 3F, 2–2–1 Hirakawacho, Chiyoda-ku, Tokyo 102, Japan

Chapman & Hall Australia, 102 Dodds Street, South Melbourne, Victoria 3205, Australia

Chapman & Hall India, R. Seshadri, 32 Second Main Road, CIT East, Madras 600 035, India

First edition 1997

© 1997 R.J. Henry

Typeset in 10/12 Sabon by Florencetype Ltd, Stoodleigh, Devon

Printed in Great Britain by Cambridge University Press

ISBN 0 412 73210 6 (HB)   0 412 73220 3 (PB)

A catalogue record for this book is available from the British Library.

Library of Congress Catalog Card Number: 97–65830

∞ Printed on permanent acid-free text paper, manufactured in accordance with ANSI/NISO Z39.48–1992 and ANSI/NISO Z39.48–1984 (Permanence of Paper).

To my parents

Contents

Preface

The application of molecular biology to the analysis and manipulation of plant genomes provides practical approaches to enhancement of the efficiency of agriculture, forestry and food production by improvement of both the quantity and quality of production. These techniques are also useful for the study and management of wild plant populations for species conservation and the control of undesirable plants (weeds). Much has been made of the potential of biotechnology to impact upon agriculture and food production, resulting in frequent expression of disillusionment by those with inflated expectations of the scope and rate of application of this technology. This book aims to address this issue by clearly identifying and illustrating the current practical possibilities and limitations of the technology, in the hope that this will allow a more realistic appreciation of the potential of this vital technology. This is an important perspective for scientists and the wider community. Appropriate application of plant molecular biology will increasingly play a crucial role in ensuring an adequate world food supply and minimizing adverse effects of agriculture on the environment.

This book outlines the current available techniques and their practical applications. Details of useful protocols are included, but the emphasis is on the identification and definition of the major opportunities to apply molecular biology in the agricultural and plant sciences. Wherever possible, specific examples of practical applications are provided to illustrate the potential uses of the technique.

The structure of the book aims to assist in the identification of opportunities to make use of plant molecular biology and provides leads to the protocols necessary for successful application. Chapter 1 describes the main molecular marker techniques available for use in plant identification. The application of these and other approaches to estimation of genetic variation in plants is covered in Chapter 2. How these techniques are applied to plant improvement is outlined in Chapter 3, while Chapter 4 introduces plant genetic engineering, including a description both of the techniques available and the possible applications. Technically, the application of molecular biology to plants is often considerably more difficult than corresponding experiments with animals and other organisms. Plants not only have thick cell walls and vacuoles in which are to be found a wide range of secondary metabolites, but also contain polysaccharides such as starch that have solubility properties similar to those of DNA. This can make the isolation of large quantities of pure nucleic acids from plants much more difficult. Useful protocols in plant molecular biology are collected in Chapter 5.

Taken together, these techniques allow plant molecular biology to be applied to practical questions such as:

- What plant (or plant-derived product) are we dealing with?
- How are these plants related?
- Which plants or plant genetic resources should receive priority for conservation?
- How can we breed or genetically engineer a plant with improved performance or more useful properties?
- What methods are available to tackle the above questions?

Key terms are listed at the end of each chapter, together with questions which are aimed to test the knowledge and understanding gained from the chapter text. Additional valuable information is included in the Appendices.

Robert J. Henry
New South Wales, Australia
1st May, 1997

Acknowledgements

I thank colleagues for giving permission for the reproduction of figures and photographs. All figures other than those listed below have been prepared especially for this book.

Figures were kindly provided by the Australian Biotechnology Association (Figure 1.2 is reproduced from *Australasian Biotechnology* and with permission from Bill Taylor, *Plant Molecular Biology*, **23**, 1011; reproduced by permission of Kluwer Academic Publishers), Ralf Dietzgen (Figure 1.3 reproduced by permission of *The American Phytopathological Society*), Wendy Lawson (Figures 1.4 and 1.17 reproduced by permission from *Australian Journal of Agricultural Research*), Brant Bassam (Figure 1.5), Narelle Egan (Figure 1.7), Life Technologies Inc. (Figures 1.8 and 1.9 reproduced with permission from *Focus*, **17**, 66–70), Peter Shewry (Figure 1.12 reproduced by permission of Cambridge University Press), Song Weining (Figure 1.13), Lien Ko (Figure 1.14 reproduced by permission of National Institute of Agricultural Botany), Glenn Graham (Figure 1.18), Boehringer-Mannheim (Figure 1.18(c)), Perkin-Elmer Corporation (Figure 1.19), David Poulsen and Bob Rees (Figure 1.22), (Figures 1.24, 1.25, 1.26 and 1.27 reproduced by permission of Academic Press), Craig Morris (Figure 1.29), Mike Smith (Figure 1.32), Glenn Graham (Figure 2.3 reproduced by permission of the *Australian Journal of Systematic Botany*), International Society of Plant Molecular Biology (Figures 2.4 and 2.5), Rod Peakall (Figures 2.6 and 2.7 reproduced by permission of Springer-Verlag), Leon Scot (Figure 2.9), Song Weining (Figure 2.12 reproduced by permission of Kluwer Academic Publishers), (Figure 3.3 reproduced by permission of Cold Spring Harbor Laboratory Press), Peter Portman and the Chief Executive Officer of the Western Australian Department of Agriculture (Figure 3.4 reproduced from the *Western Australia Journal of Agriculture*), Nick Tinker and Diane Mather (Figure 3.5), Wendy Lawson (Figure 3.6), Olin Anderson (Figure 4.2), Freek Heidekamp (Figure 4.8) and Kutty Kartha (Figure 4.16), all reproduced with permission of Plenum Publishing Corporation), Mahin Abedinia (Figure 5.3), Stratagene (Figure 5.5), Stockton Press (Figure 5.6), Clontech (Figure 5.7), New England Biolabs (Figure 5.8), Clontech (Figures 5.11 and 5.12), and Gay McKinnon (Figures 5.13, 5.14 and 5.15). Megan Jones, Ron Marschke and Gordon Guymer also provided figures.

I thank my colleagues, especially those at the Queensland Agricultural Biotechnology Centre for encouraging me to embark upon the preparation of this book. I also thank Margot Henry for continuing unlimited support in this project. Peta Carolan provided essential assistance in the assembly of the manuscript.

Field testing of new wheat varieties (Trangie, New South Wales, Australia). Molecular biology provides tools for developing more efficient and sustainable agriculture and food production.

Identification of plants using molecular techniques | 1

Identification of plants from wild populations is one practical application of plant molecular biology (the photograph shows a forest area in Japan north of Tokyo). However, the same techniques also find use in agricultural and food industries.

After reading this chapter you should understand:

● Molecular Techniques ● Restriction Fragment Length Polymorphism ● Random Amplified Polymorphic DNA ● Short Sequence Repeats ● Amplified Fragment Length Polymorphism ● Ribosomal Gene Analysis ● Sequence Tagged Sites ● Methods for Detection of PCR Products ● Heteroduplex Analysis ● Protection of Plant Breeders' Rights ● Principals of Plant Variety Protection ● Techniques for Variety Protection● Identification of Plant Pathogens ● Need for Pathogen Identification ● Techniques for Pathogen Identification ● Commercial and Industrial Applications ● Species Identification ● Variety Identification ● Somaclonal Variants

1.1.1 INTRODUCTION

Probably the most immediate practical application of the techniques of molecular biology to plants is in plant identification. Determining or confirming the identity of a plant or plant derived product is usually the first step in its study or use.

Other applications of molecular biology, such as the development of new plant varieties by transformation or marker-assisted breeding, may take years while positive identification of a plant species or variety may be possible within one day or in some cases much sooner.

The techniques that are used for plant identification are designed to detect the presence of specific DNA sequences or combinations of sequences that uniquely identify the plant. However, this does not usually require DNA sequencing but is generally based upon either nucleic acid hybridization or the polymerase chain reaction (PCR). Techniques based upon hybridization, such as restriction fragment length polymorphism, have been available for many years and continue to be improved. The polymerase chain reaction (Saiki *et al.*, 1988) (Figure 1.1) has been the basis of a growing range of newer techniques (Table 1.1). The PCR-based tests generally have advantages in speed and sensitivity but in some cases hybridization based tests may be cheaper to conduct.

The choice of method for any particular application will depend upon the difficulty of the distinction to be made and factors such as the time, facilities and funds available. None of the currently available methods will be the best option in all cases and successful practical application of the techniques of molecular biology to plant identification requires flexibility in the choice of protocols. The nature of the sample may determine the techniques that are able to be employed.

Polymerase chain reaction (PCR)

The polymerase chain reaction allows the specific amplification of DNA sequences, making it ideal for the identification of plant genotypes. Amplification of a genotype-specific sequence can take advantage of some of the many features of PCR.

- **Speed:** rapid cyclers allow amplification in a few hours in the slowest cases, but may be very much faster.
- **Simplicity:** a single reaction mix can contain all of the reagents necessary for specific amplification of diagnostic sequences. A single solution can be added to a DNA sample and subjected to temperature cycling.
- **Specificity:** the specificity can be adjusted to the level required, family, genus, species or individual genotype by the choice of primers.
- **Sensitivity:** the ultimate sensitivity of the PCR can approach single molecule detection, giving sensitivity at the theoretical limit.
- **Cost:** the cost of the consumed reagents may be kept low by conducting the PCR in a small volume.

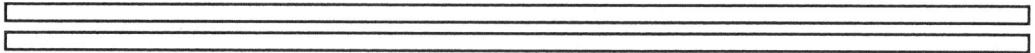

Denaturation

⇓

Primer annealing

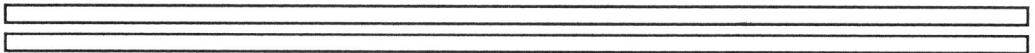

⇓

Chain elongation

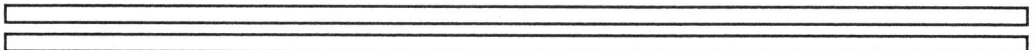

Table 1.1 PCR-based molecular marker methods for plant identification

RAPD	Random amplified polymorphic DNA
AP-PCR	Arbitrarily primed PCR
DAF	DNA amplification fingerprinting
SSR	Simple (short) sequence repeat
STR	Short tandem repeat
AFLP	Amplification fragment length polymorphism
STS	Sequence tagged site
SCAR	Sequence characterized amplified region

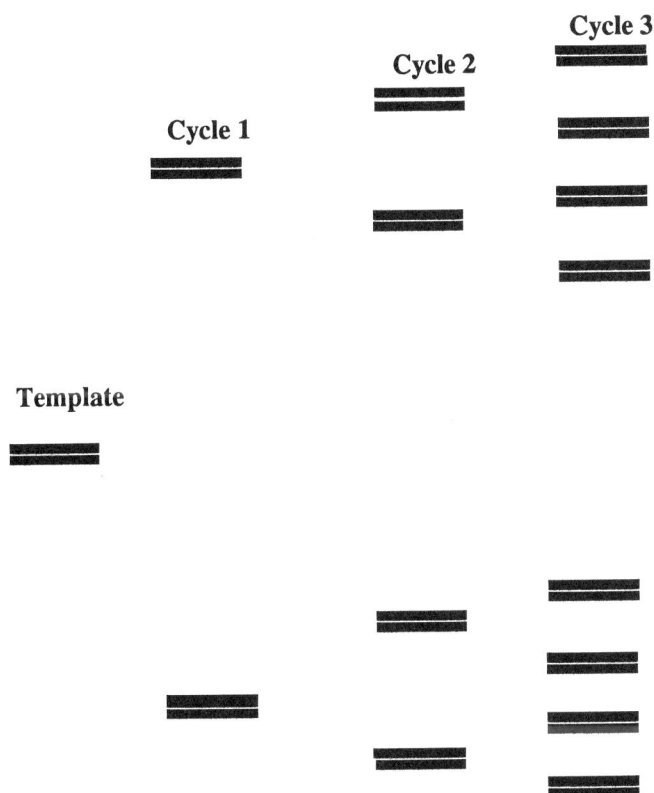

Figure 1.1 Polymerase chain reaction (PCR). (a) The process requires repeated cycles of denaturation, primer annealing and primer extension leading to amplification of the target sequence. (b) Note 20 cycles provides a one million-fold amplification at 100% efficiency. For more background see Mullis (1990).

The size of the sample available may be too small for some techniques. The age and conditions of storage of the sample will also limit the options. Some techniques require a careful isolation of large amounts of highly purified DNA while others can be performed on crude extracts of the plant sample. Protocols for the preparation of samples for analysis of DNA are described in Chapter 5.

The identification of plants from a specific plant variety or species depends upon the definition of distinguishing genetic differences between the plant to be identified and other plants. Analysis of genetic relationships between plants is described in Chapter 2. The analysis of these differences may be an essential prerequisite to the identification of plants using molecular approaches. The distinguishing sequences revealed in such studies may be exploited to develop diagnostic tests for specific species or varieties of plants. This chapter will outline the options for plant identification using molecular approaches and indicate the possible applications of plant identification based upon molecular methods.

1.1.2 RESTRICTION FRAGMENT LENGTH POLYMORPHISM (RFLP) AND OTHER HYBRIDIZATION-BASED METHODS

Restriction fragment length polymorphism (RFLP) is based upon hybridization of a probe (a probe is a specific DNA sequence designed to hybridize with and thus detect a target sequence or sequences in the unknown sample) to fragments of genomic DNA following digestion with restriction enzymes. Differences in the sequence at or around the sequence with which the probe hybridizes may result in differences (polymorphisms) in the length of the fragments detected by the probe (Figure 1.2). Genomic DNA from the sample being tested is digested with a restriction endonuclease. These enzymes cleave DNA at specific sites with sequences (four or more base pairs) recognized by the enzyme. The resulting DNA fragments are separated by electrophoresis on an agarose gel and transferred by blotting onto a nylon membrane to allow hybridization with a probe. The DNA is crosslinked to the membrane and then hybridized to the pre-labelled probe under conditions of appropriate stringency. Detection of the probe follows with the protocol depending upon the type of label used. This technique requires the availability of a suitable DNA probe, usually from the species being studied. Detection of single copy sequences in larger plant genomes may require the isolation of relatively large amounts of DNA (10 µg or more) in a form that is able to be digested with the restriction enzyme. Related species may also be useful sources of probes. The main types of probes are cDNA or genomic clones. Conserved and multiple copy genes such as ribosomal RNA genes are useful general probes for this type of analysis.

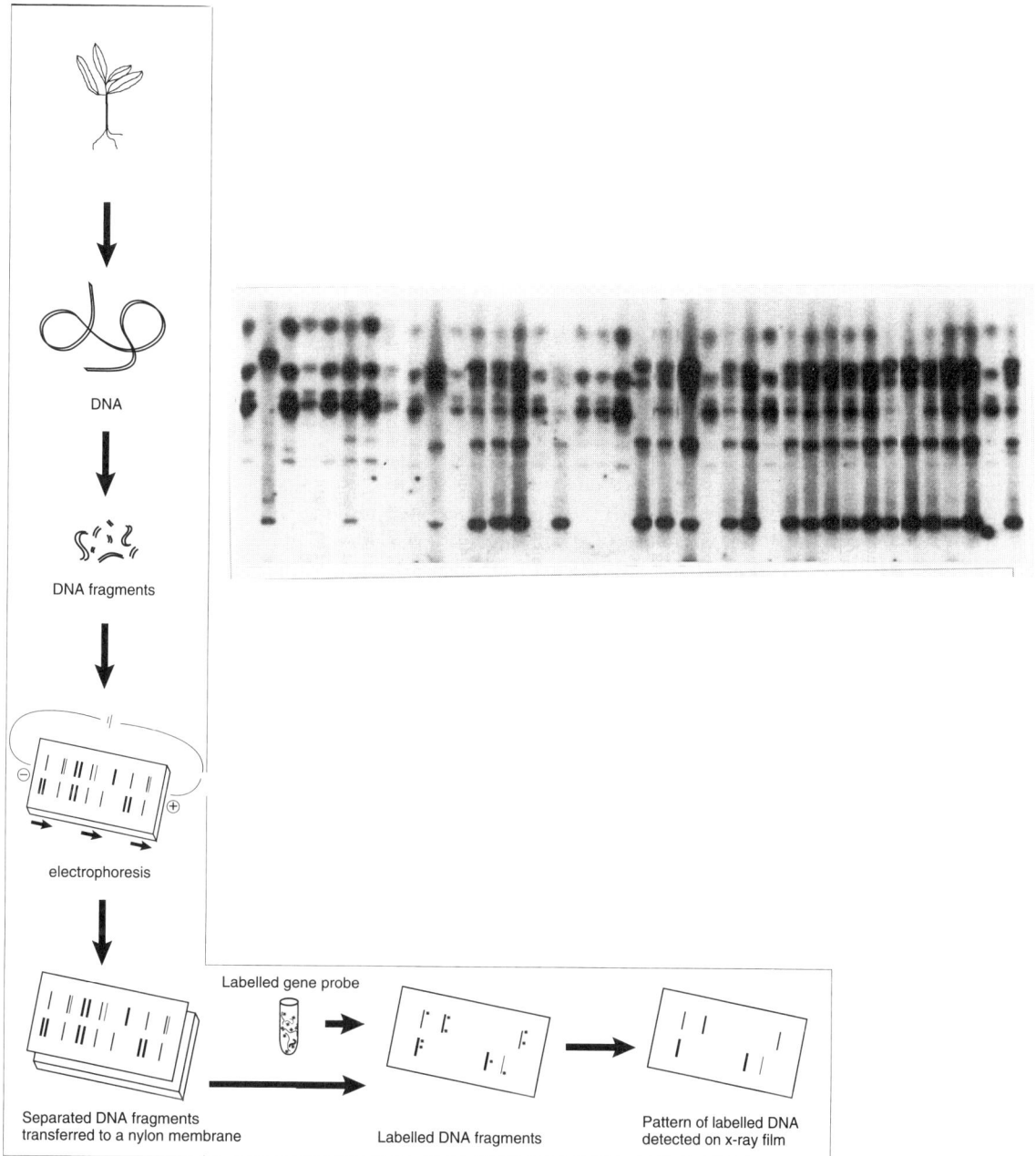

Figure 1.2 Restriction fragment length polymorphism. The gel illustrates segregation of RFLPs in F_2 progeny of an interspecific hybrid tomato (Taylor *et al.*, 1993). DNA was digested with *Hae*III and probed with a cDNA for a proteinase inhibitor.

Table 1.2 Non-radioactive methods for the labelling of DNA

Label*	Detection method
Alkaline phosphatase	Colorimetric or chemiluminescent substrate
Biotin	Streptavidin labelled with enzyme (alkaline phosphatase)
Digoxigenin	Antibody labelled with fluorescent dye or enzyme (alkaline phosphatase)
Fluorescein	Fluorescence

*Directly attached to DNA

The labelling of probes was traditionally done with radioisotopes, thus restricting the use of the technique in some laboratories. A range of non-radioactive labelling techniques have become available with increasing sensitivity (Table 1.2).

Direct hybridization in dot or slot blots may also be used for the detection of specific sequences that identify the presence of a specific plant variety or species (Figure 1.3). In these procedures, samples of the test DNA are applied directly to membranes and probed. Species-specific DNA sequences can be identified by cloning of repeated sequences (Guidet *et al.*, 1991) and these clones may be used to detect the presence of the species.

Repeated sequences [usually 9–65 base pairs (bp) long] that occur a variable number of times (10–300 times) may provide a very high level of polymorphism detectable by hybridization. This special case of RFLP analysis has been termed VNTR (variable number of tandem repeats) analysis and these markers are otherwise known as mini-satellites (Gepts *et al.*, 1992). VNTR markers have been used to distinguish the genotype of plants such as rice (*Oryza sativa*) (Winberg *et al.*, 1993) and bean (*Phaseolus vulgaris*) (Stockton *et al.*, 1992). The application of minisatellite analysis in plants has been limited but development of PCR-based methods for detection of these polymorphisms (limited by the large size of the fragments that would need to be amplified) may increase the level of adoption. These markers have been studied in turnip (*Brassica rapa*) using probes generated by PCR (Rogstad, 1994).

1.1.3 RANDOM AMPLIFIED POLYMORPHIC DNA (RAPD)

A widely applied approach to characterization of DNA from plants and other organisms is to use PCR with short oligonucleotide primers of arbitrary (random) sequence to generate genetic markers. This is the basis of the random amplified polymorphic DNA (RAPD) method

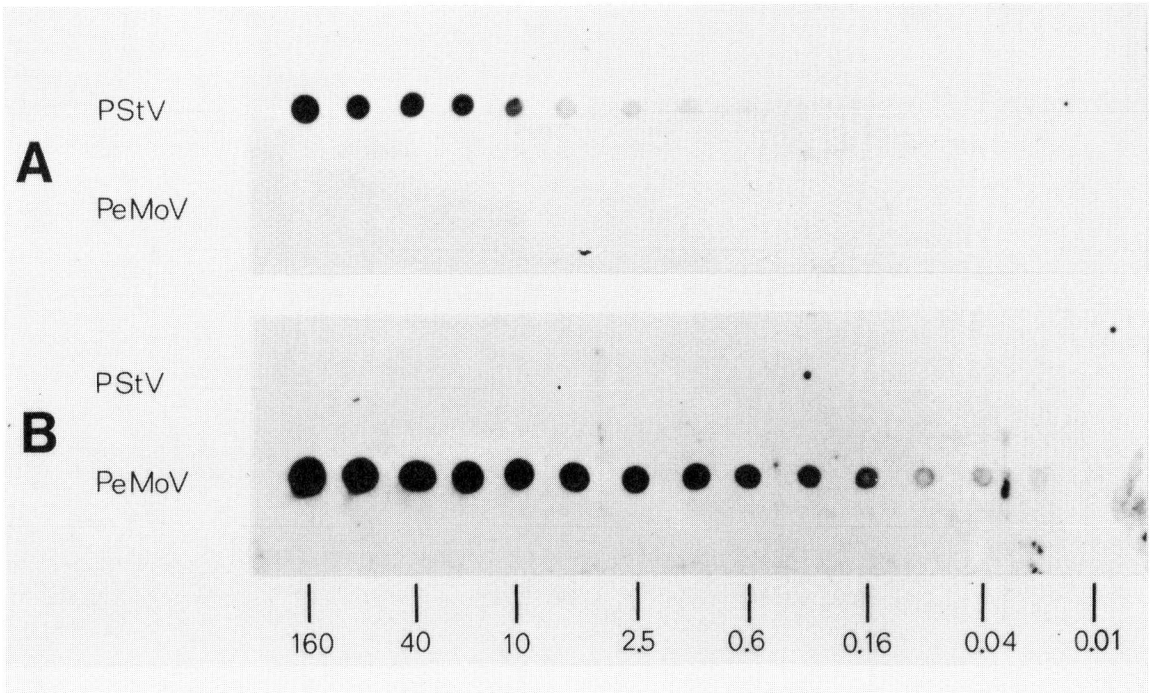

Figure 1.3 Dot blot detection of specific sequences. Two peanut viruses (peanut mottle and peanut stripe) are distinguished using specific probes (Dietzgen *et al.*, 1994). A two-fold dilution series of peanut stripe (PStV) and peanut mottle (PeMoV) were hybridized with cRNA probes (A) PStV or (B) PeMoV. The scale indicates the amount of the virus (ng). Note the high specificity but different sensitivity of the two probes.

(Williams *et al.*, 1990), arbitrarily primed polymerase chain reaction (AP–PCR) (Welsh and McClelland, 1990) and DNA amplification fingerprinting (DAF) (Caetano-Anolles *et al.*, 1991).

The RAPD technique has been adopted most widely. The main issues associated with the use of these techniques is the problem of ensuring reproducibility of amplification profiles. The nature of the amplification process with short primers is such that many sites in the genome are potential templates and the profile obtained may be influenced by any variation in the method used to prepare the DNA template and the exact reaction composition and conditions used in the PCR. This means that variation in the concentration of, for example, the primer or template can result in the amplification of different products (Muralidharan and Wakeland, 1993). Obtaining reliable results depends upon standardizing these conditions or identifying combinations of conditions that give consistent results, even when variations in the key variables are encountered. Standard primer, nucleotide and magnesium concentrations,

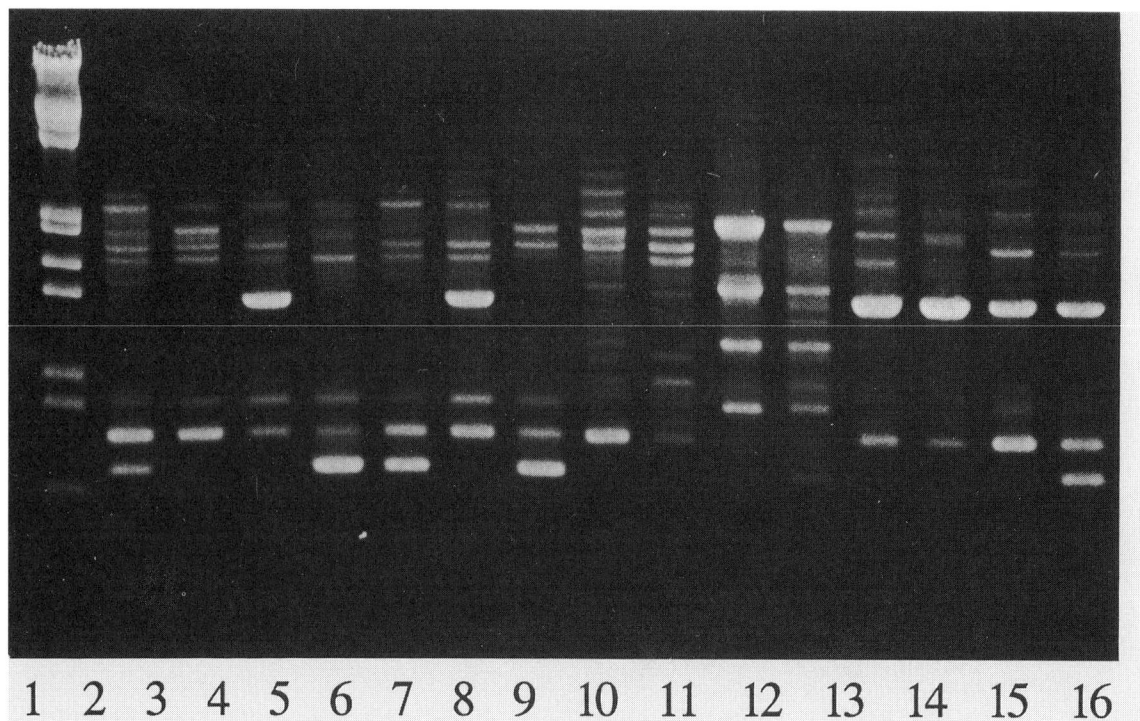

Figure 1.4 Random amplified polymorphic DNA (RAPD). Lane 1, molecular size marker (Lambda- *Hind*III/*Eco*RI); lanes 2–16, RAPD–PCR of 15 sunflower varieties with the primer 5'-GTGTGCCCCA-3' (Lawson *et al.*, 1994). Note the high level of polymorphism between sunflower varieties.

exact reproduction of temperature cycling conditions and DNA polymerase type and activity are essential. However, a key requirement for reliable and reproducible RAPD results is a consistent approach to sample preparation and DNA isolation. Both the quantity and the quality of the template DNA preparation have the potential to substantially influence the result. Protocols for arbitrarily primed PCR (RAPD and DAF) are given in Chapter 5.

The RAPD method makes use of agarose gels to analyse the PCR products. RAPD results for commercial sunflower varieties are shown in Figure 1.4. This technique is capable of identifying variation within cultivars. However, even in variable outcrossing species such as rye, RAPD analysis can be used to distinguish varieties (Iqbal and Rayburn, 1994).

The DAF techniques is based upon polyacrylamide gels and silver staining. DAF results for fungal isolates are given in Figure 1.5. The DAF technique generally generates more scorable bands because of the higher resolution of the gels, but the agarose gel separations used with the RAPD method are simpler to perform. A comparison of the

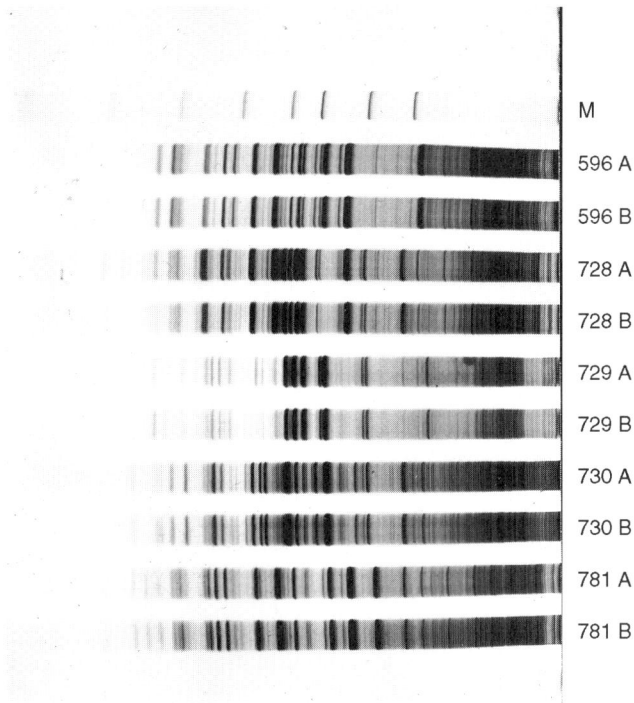

Figure 1.5 DNA amplification fingerprinting (DAF). Profiles from the fungus *Pythium* isolated from sugarcane soils. Note the reproducibility of the duplicates (separate DNA isolation and amplification) and the ability to distinguish between and within species. Lanes from left to right, markers (from top 1000, 700, 500, 400, 300, 200, 100, 50 bp), *Pythium arrhenomanes* isolate UQ596 two lanes, isolate UQ728 two lanes, *P. graminicola* two lanes, *P. spinosum* two lanes, *P. arrhenomanes* isolate UQ781 two lanes.

main features of the different arbitrary primer methods is given in Table 1.3

The RAPD method may reveal more polymorphisms if combined with restriction digestion. Wheat is a species with little genetic variation and digestion of genomic DNA from wheat with restriction enzymes before RAPD analysis has been shown to reveal more polymorphisms (Riede *et al.*, 1994).

1.1.4 SIMPLE SEQUENCE REPEATS (SSR)

Simple repeat sequences, di- or tri-nucleotide repeats are common in plant genomes. Simple (or short) sequence repeats (SSRs) are also known as STRs (short tandem repeats) or 'microsatellites'. PCR using

Table 1.3 Comparison of arbitrary primer methods (Bassam and Bentley, 1994)

	RAPD	AP-PCR	DAF
Resolution	Low	Intermediate	High
Products	1–10	3–50	10–100
Separation	Agarose	Polyacrylamide	Polyacrylamide
Visualization	Ethidium bromide	Radiolabelling	Silver
Primer length*	9–10	20–35	5–15
Primer concentration	0.3 μM	3 μM	3–30 μM

*Nucleotides.

primers to the sequences flanking these repeats can be used to generate polymorphisms because of frequent variation in the length of the repeat region (Figure 1.6). Microsatellites have been established as useful genetic markers in many plant species (Wu and Tanksley, 1993; Zhao and Kochert, 1993; Saghai Maroof et al., 1994; Roder et al., 1995). Another approach is to amplify the regions between SSRs, in inter-SSR–PCR (Zietkiewicz et al., 1994).

The main limitation on the application of this technique in plant identification has been the difficulty of cloning and sequencing the regions flanking the SSR. Considerable effort has been devoted to the development of efficient protocols for the cloning and sequencing of SSRs (Taylor et al., 1992). This must be done for each species and even then the PCR primers designed following cloning and sequencing may not always reveal a high level of polymorphism. The flanking regions are relatively species specific. Markers developed for a particular species are not usually useful for application to even closely related species. The reliability and reproducibility of the markers, especially between laboratories, makes them attractive alternatives to techniques such as RAPD–PCR when they have been established for the species under investigation. Analysis of plant gene sequences in databases suggests that microsatellites are less abundant than in mammals (Lagercrantz et al., 1993) and should be found on average once every 50 kb throughout the plant genome making them very useful as genetic

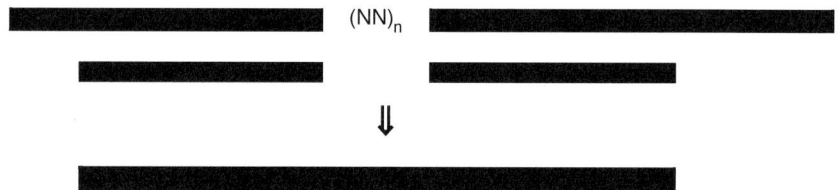

Figure 1.6 PCR analysis of variations in the length of short sequence repeats (SSR). A dinucleotide repeat is depicted.

Repeated sequences and 'satellite' DNA

Tandemly repeated sequences are found in most eukaryotic organisms, including plants. DNA with long repeats (100–1000s) has been called 'satellite' DNA because it was first detected as a distinct band separated from the main genomic DNA band in equilibrium density centrifugation. Tandem repeats with shorter repeat units (10–100) were then called minisatellites and very short repeat units (1–4) micro-satellites.

---

markers (Morgante and Olivieri, 1993). The frequency of repeat sequences differs from that in animals with AT repeats being the most common dinucleotide repeat. The SSR sequences are mostly found in introns and the 5' flanking region of plant genes (Table 1.4). An example of PCR of an SSR from wheat is shown in Figure 1.7. Roder *et al.* (1995) estimated that a haploid wheat genome contained 3.6×10^4 blocks of $(GA)_n$ and 2.3×10^4 blocks of $(GT)_n$. The estimates suggest that these two microsatellite repeats alone provide a marker even 270 kb on average. Dinucleotides in wheat had up to 40 repeats. A wide range of other microsatellite repeat motifs were also detected.

PCR with primers complementary to the SSR regions themselves can also be used to generate marker bands. These markers could be used in a similar way to RAPDs (Weising *et al.*, 1995). Many of the bands generated using this approach, apparently result from some degree of mismatching.

1.1.5 AMPLIFICATION FRAGMENT LENGTH POLYMORPHISM (AFLP)

Amplification fragment length polymorphism (AFLP) (Zabeau and Vos, 1993) is a method for PCR amplification of restriction digests of

Table 1.4 Distribution of microsatellite repeat sequences (SSRs). (After Morgante and Olivieri, 1993.)

	No. of microsatellites identified in search of sequence database					
	5' flank	*Introns*	*3' flank*	*5' untranslated*	*3' untranslated*	*Coding*
Dinucleotide	23	22	8	2	13	–
AT	21	14	8	–	7	–
AG/TC	2	7	–	2	6	–
AC/TG	–	1	–	–	–	–
Trinucleotide	7	6	2	8	3	12

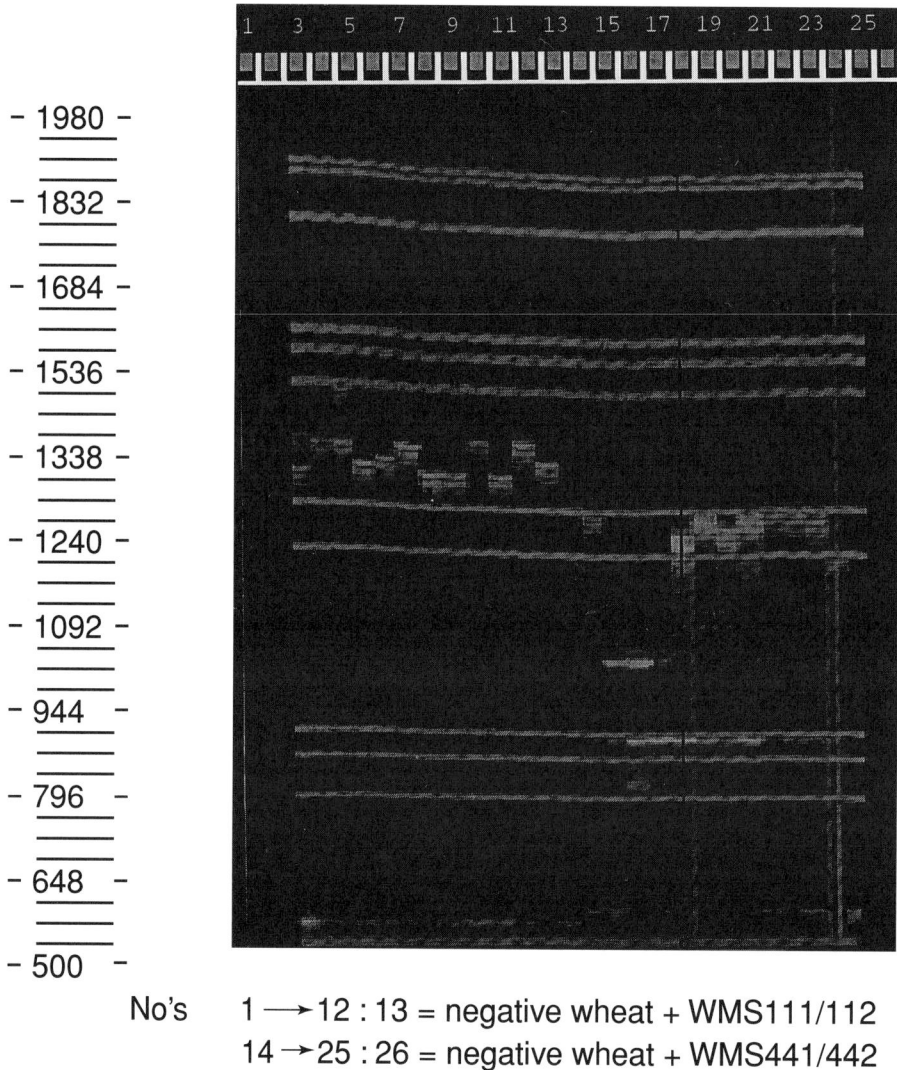

No's   1 ⟶ 12 : 13 = negative wheat + WMS111/112
        14 ⟶ 25 : 26 = negative wheat + WMS441/442

Figure 1.7(a)

genomic DNA following the ligation of oligonucleotides (Figure 1.8). DNA from the plant to be analysed is digested with a restriction enzyme. Short oligonucleotides are then ligated to the ends of all fragments. A subset of the restriction fragments is then amplified by PCR with primers complementary to the added oligonucleotide and restriction site. Additional specificity is provided by a small number of specific nucleotides added at the 3' end of the PCR primer. Each base added at the 3' end of the primer reduces the number of fragments

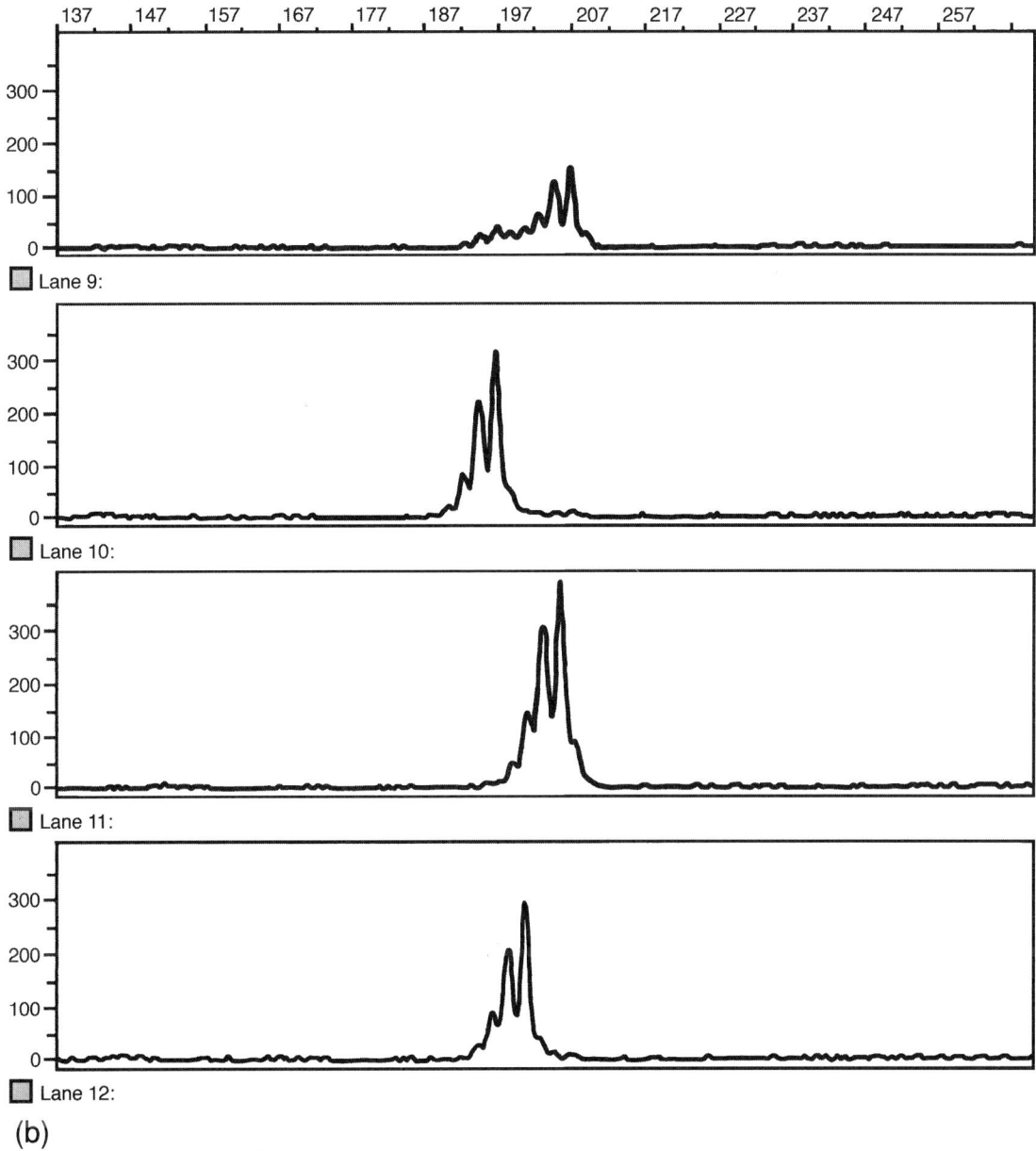

Figure 1.7 Polymorphism in short sequence repeats in wheat. (a) Lanes 1–12 and 14–25 represent two different sets of microsatellite primers. (b) Profiles for lanes 9, 10, 11 and 12 from the cultivars Rosella, Suneca, Janz and Halberd are shown. Note the characteristic stutter bands amplified for these dinucleotide repeats.

Restriction digestion   ⇩

Ligate oligonucleotides   ⇩

Anneal primers   ⇩

Amplify   ⇩

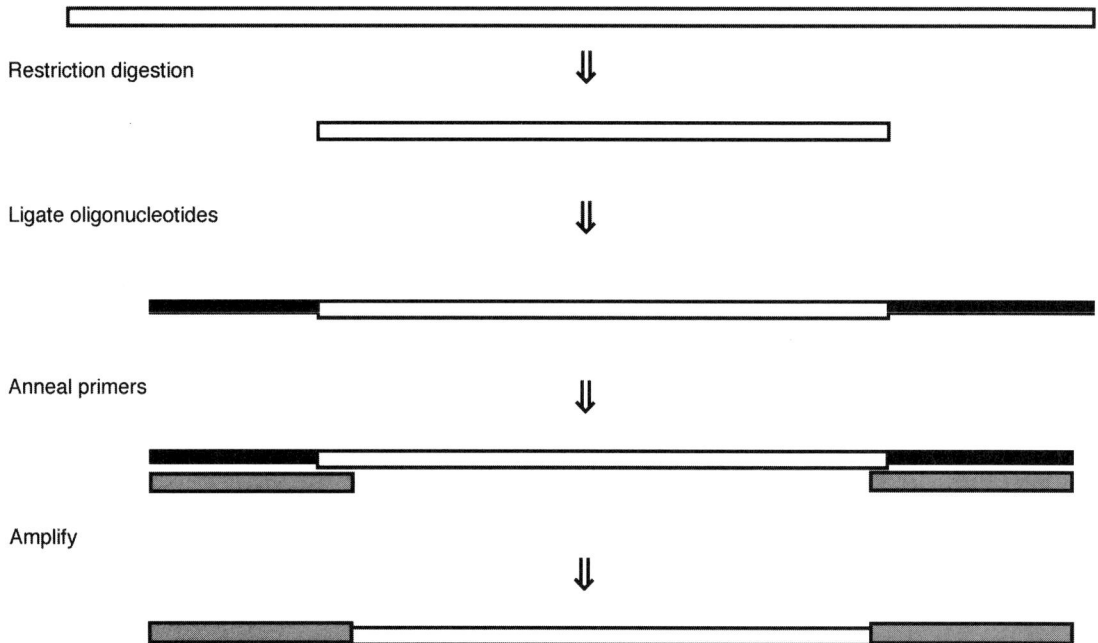

Figure 1.8 Amplified Fragment Length Polymorphism (AFLP) depicted schematically

amplified by a factor of 16 on average because of the four-fold selection at each end. An example of AFLP analysis is given in Figure 1.9.

The AFLP technique generates large numbers of molecular markers, making the use of automated gel analysis highly desirable. Labelling with fluorescent dyes and analysis using automated sequencing equipment has been adopted for this technique. In a simpler form, AFLP markers may also be used to identify related clones in gene libraries (e.g. cosmids, YACs or BACs; Chapter 5) by using no or very few selective nucleotides in the PCR amplification (Vos *et al.*, 1995).

1.1.6 RIBOSOMAL GENE ANALYSIS

The ribosomal genes of plants are attractive targets for molecular analysis because of the presence of large numbers of copies of these genes in the genome. PCR of these genes provides a very sensitive test. The 5S RNA genes are highly conserved. Primers to these genes can be used to amplify the spacers between the 5S RNA genes (Kolchinsky *et al.*, 1991; Ko *et al.*, 1994) (Figure 1.10). The length and sequence of these spacers is characteristic of the species. Eukaryotic 18S–5.8S–25S

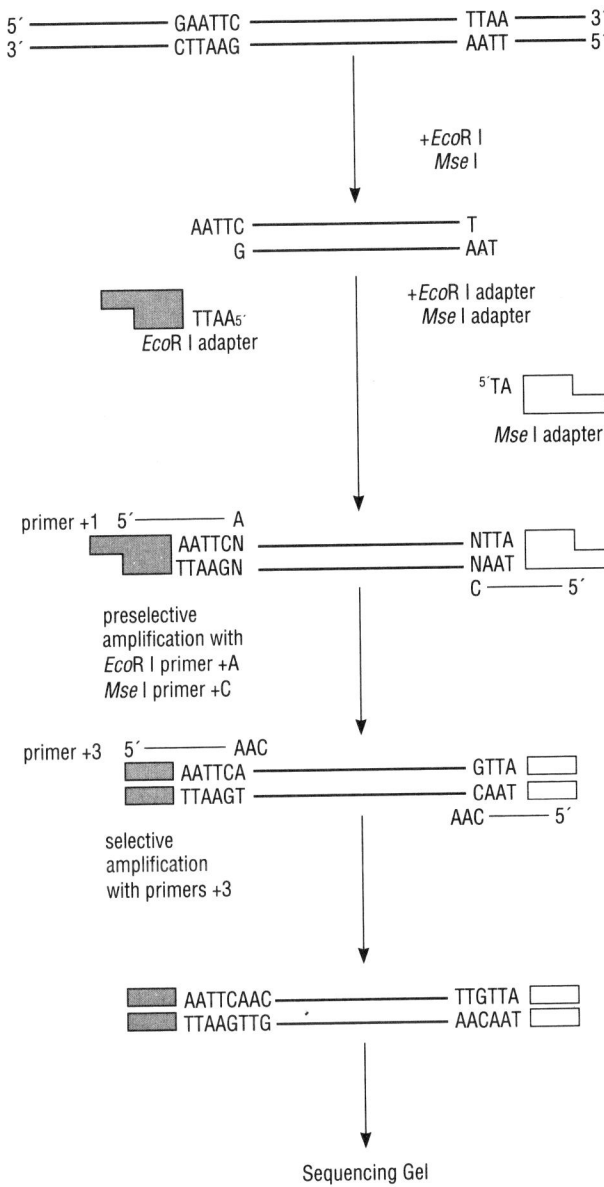

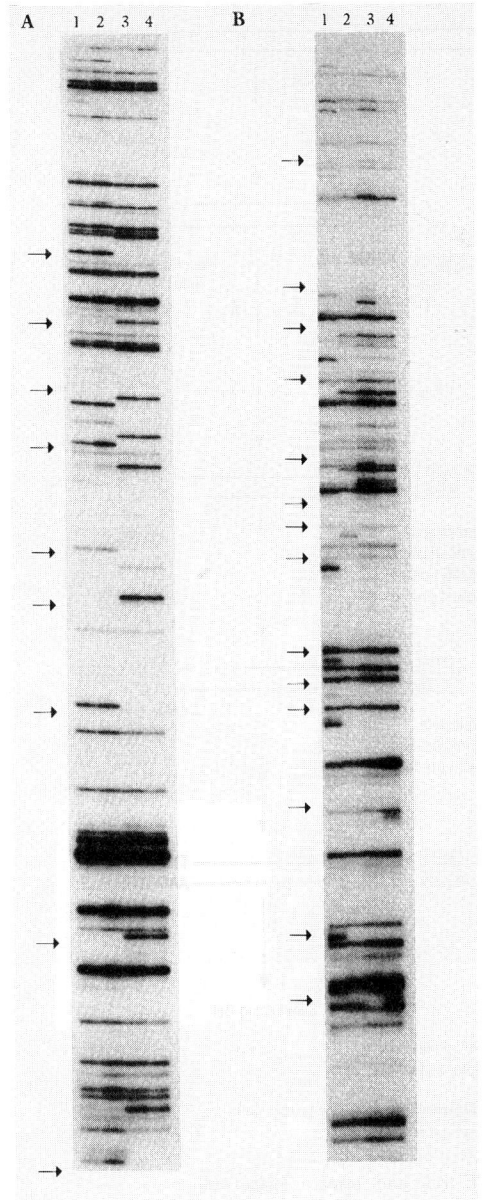

Figure 1.9 Examples of AFLP analysis. (a) Specific example with *Mse*I and *Eco*RI adapters [Lin, J.-J. and Kuo, J. (1995) *Focus*, **17**, 67]. (b) A, *Arabidopsis thalianas* ecotypes; B, *E. coli* strains [Lin, J.-J. and Kuo, J. (1995) *Focus*, **17**, 68].

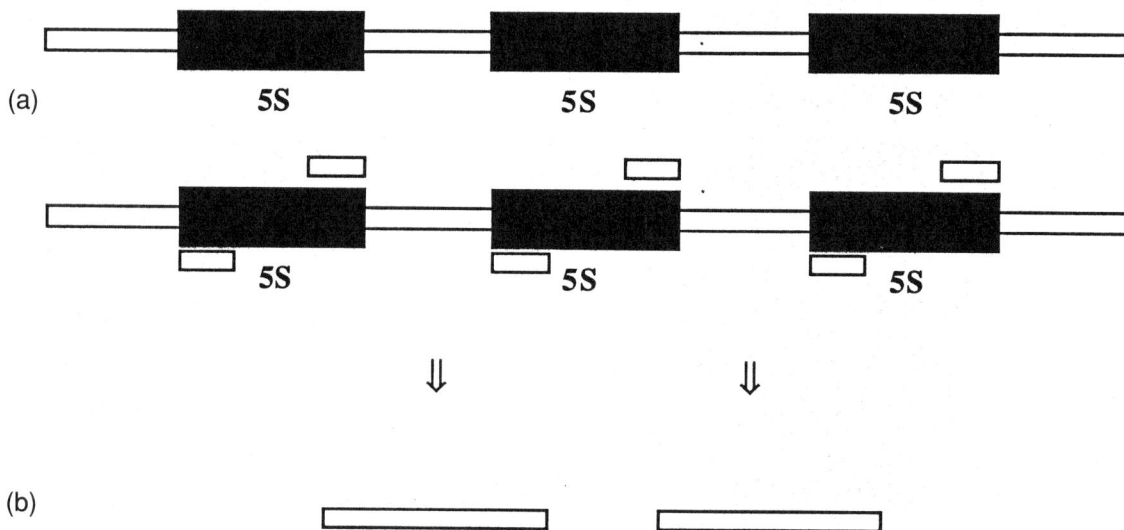

(a)

(b)

Figure 1.10 5S ribosomal RNA genes. (a) Arrangement of genes and (b) use of PCR for spacer amplification

ribosomal RNA (Figure 1.11) is transcribed in a single unit with an intergenic spacer (IGS). The three mature rRNAs are produced by processing this transcript removing the two internal transcribed spacer (ITS) sequences. The ITS and IGS regions are both possible sources of polymorphism for plant identification.

1.1.7 STORAGE PROTEIN GENE ANALYSIS

Seed storage proteins are encoded by gene families that offer polymorphism with potential for exploitation in plant variety identification. The cereal storage proteins have been used in this way (Figure 1.12). PCR of the high molecular weight (HMW) and low molecular weight (LMW) glutenin subunits of bread wheat (D'Ovidio *et al.*, 1992; Kolster *et al.*,

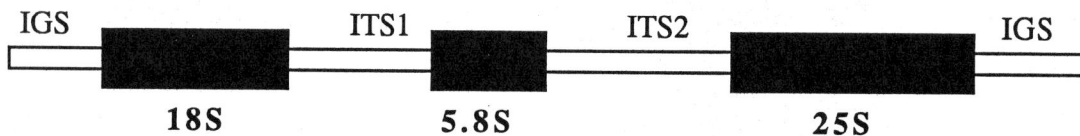

Figure 1.11 Ribosomal RNA genes of eukaryotes: arrangement of the 18S–5.8S–25S ribosomal RNA gene complex. IGS, intergenic spacer; ITS, internal transcribed spacer.

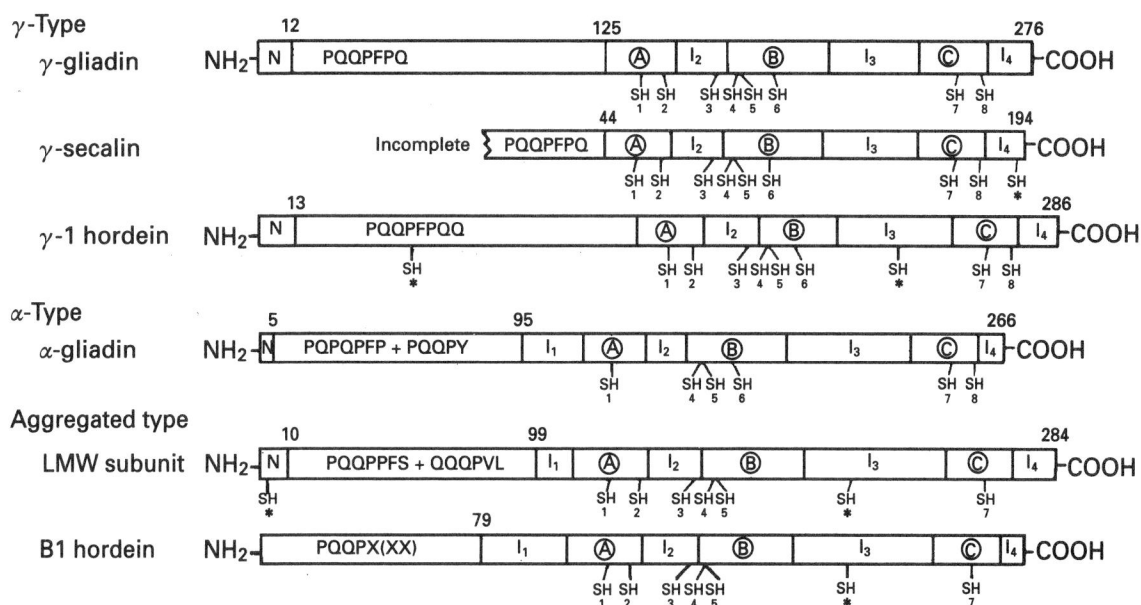

Figure 1.12 Seed storage proteins. Schematic structures of the S-rich prolamines of the grasses (Tritiaceae) as depicted by Shewry (1995). Note the conserved (encircled A, B and C), intervening (I) and repeat regions providing options for PCR-based analysis.

1993) and PCR of gliadin genes in durum wheats (D'Ovidio, 1993) have been suggested as methods for distinguishing varieties.

The repetitive sequences encoding the long runs of amino acid repeats in these storage proteins represent microsatellites. Devos *et al.* (1995) have characterized microsatellite sequences in wheat storage proteins (Table 1.5).

Sunflower storage protein genes have been used to identify sunflower varieties (Brunel, 1994). The 2S albumin gene family encodes small polypeptides that represent more than 50% of the seed storage protein

Table 1.5 Microsatellites in wheat genes. (After Devos *et al.*, 1995.)

Gene family	Microsatellite	Location
α-amylase 3	$(TA)_{12}CA(TA)_9$	Upstream of coding region
α-amylase 2	$(GCT)_7$	Coding region
α,β-gliadin	$(CAA)_n$	Coding region
γ-gliadin	$(CAG)_n(CAA)_n$	Coding region
LMW-glutenin	$(CAG)_5(CAA)_8$	Coding region

in sunflower. Primers were designed from the sequence of the HAG5 gene in a region of high melting point in the first exon. The products generated by PCR with these primers were of similar length and did not reveal any polymorphism when analysed by electrophoresis in non-denaturing conditions. Polymorphism was revealed by denaturing gradient gel electrophoresis. Techniques capable of separating DNA of the same size but with differing sequence are likely to be useful in detecting polymorphisms in storage protein and other gene families. Alternatively, the value of this approach may be enhanced by using restriction enzymes to digest the PCR products and reveal polymorphisms in storage protein genes of similar length.

1.1.8 ALPHA-AMYLASE GENE ANALYSIS

Other approaches may be based upon the detection of polymorphisms within specific genes or gene families. The cereal alpha-amylase gene family has been exploited to reveal polymorphisms for variety identification (Figures 1.13 and 1.14). Primers to many parts of the alpha-amylase genes of barley have been designed and tested for use in variety distinction in cereal species (Weining and Langridge, 1991; Ko and Henry, 1994a). These primers were designed to hybridize with highly conserved regions but amplify fragments containing potential length polymorphisms. Maize, sorghum and rice varieties can be readily distinguished using some combinations of these primers.

1.1.9 PCR OF OTHER GENES

Analysis of variations in other genes by PCR may also reveal useful polymorphisms. Denda *et al.* (1995) found length variations in intron

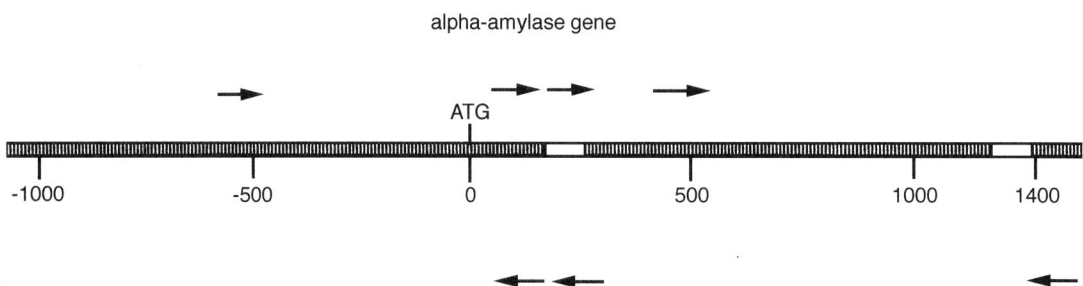

Figure 1.13 PCR of alpha-amylase genes (Ko *et al.*, 1996). Positions of primers that have been used in PCR analysis are indicated by the arrows. Numbers indicate the nucleotide number relative to the translation initiation codon.

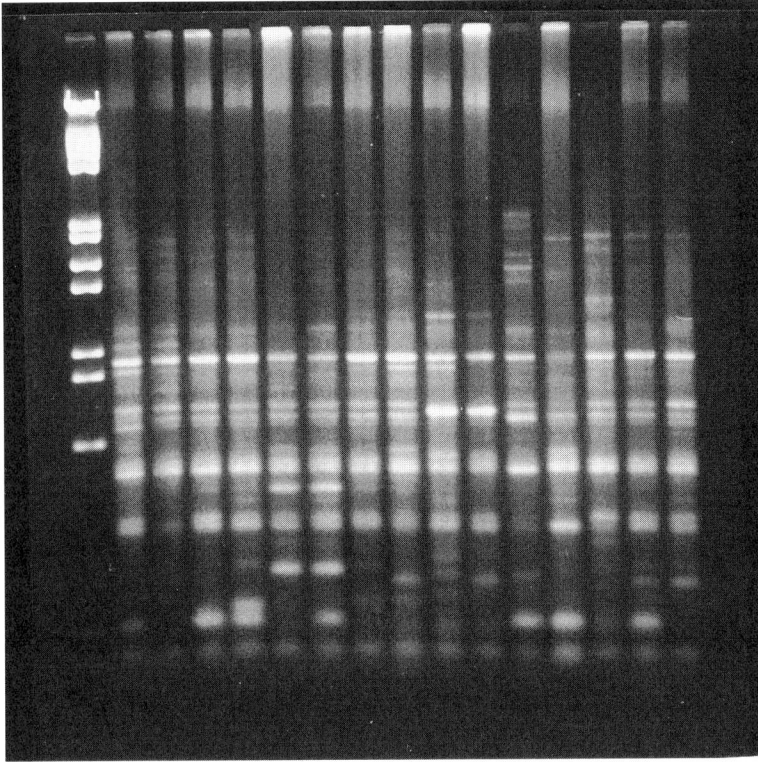

Figure 1.14 Distinction of maize varieties by PCR using alpha-amylase primers as reported by Ko *et al.* (1996). The first lane is a size marker and the remaining 15 lanes are the products of PCR with primers to α-amylase genes and DNA from different maize varieties.

3 of the *Adh* gene in the genus *Brachyscome*. Ten species gave 18 different PCR products when the third intron was amplified. The intron varied in length from 113 to 548 bp, providing easily detected polymorphism. Any gene with conserved sequences flanking regions of variable length or sequence can potentially be exploited to reveal a polymorphism that can be used as a genetic marker. PCR primers based upon the conserved sequences can be designed and used to amplify fragments distinguishable on the basis of their size or sequence.

1.1.10 INTRON SPLICE JUNCTION PRIMERS

Primers (Figure 1.15) representing a consensus of the intron splice junction (ISJ) have also been used in an attempt to target amplification

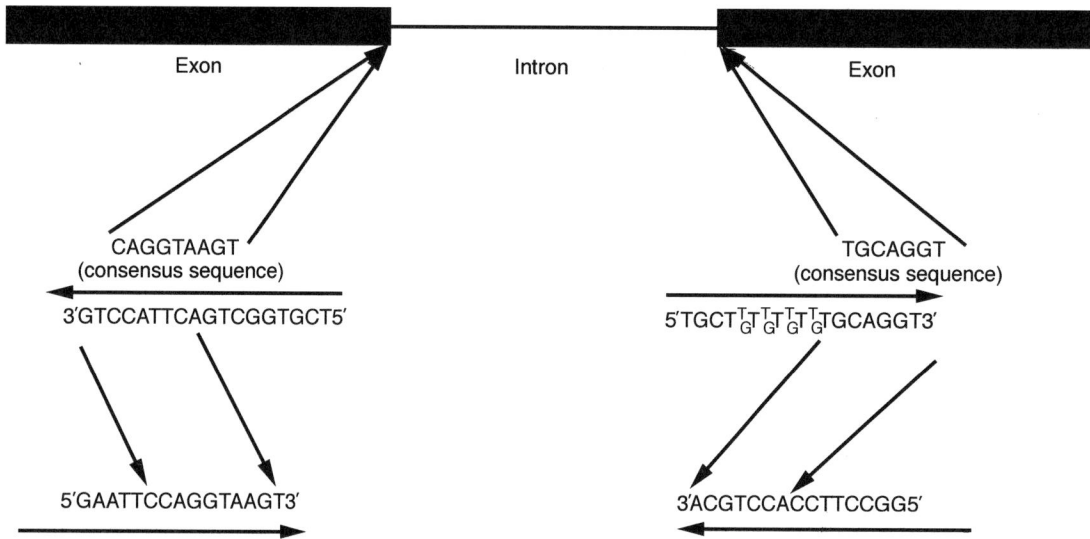

Figure 1.15 Intron splice junction primers. The design of consensus intron/exon splice junction primers is depicted schematically as described by Weining and Langridge (1991).

of coding regions (Weining and Langridge, 1991). These primers have been combined with primers of arbitrary sequence to produce polymorphic markers. This method has been successfully applied to the study of wild barley (Weining and Henry, 1995). ISJ primers can also be combined with primers for specific genes to reveal useful polymorphisms (Weining and Langridge, 1991).

1.1.11 SEQUENCE TAGGED SITE (STS) AND SEQUENCE-CHARACTERIZED AMPLIFIED REGION (SCAR)

A sequence tagged site (STS) may be based upon any known sequence in the genome. Examples from the barley genome are given in Table 1.6. These primers can be used to distinguish north American barley varieties with each primer pair revealing two or more alleles in 14 varieties tested (Chee *et al.*, 1993). Sequence-characterized amplified region (SCAR) is a term used to describe PCR methods developed by sequencing markers amplified in arbitrary primer experiments. The sequence of the amplified product is used to design longer primers that offer greater specificity (Figure 1.16). A SCAR may be applied to routine screening with simple DNA preparations without the restrictions that may have applied to obtaining reliable data from the RAPD

Sequence Characterized Amplified Region

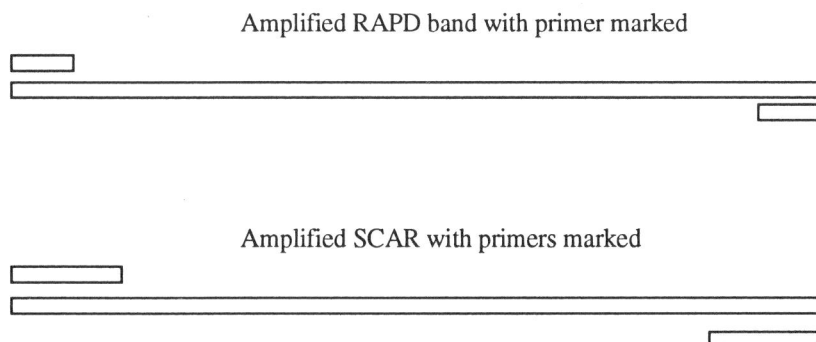

SCAR

Amplified RAPD band with primer marked

Amplified SCAR with primers marked

Figure 1.16 Sequence Characterized Amplified Region (SCAR) The sequencing of a RAPD amplification product allows the design of longer specific primers for the amplification at the same locus.

analysis on which it was based. Figure 1.17 illustrates a SCAR marker. SCAR markers may be considered as a type of STS marker.

1.1.12 SPECIFIC DETECTION OF PCR PRODUCTS

The products of any of the above PCR methods may be detected in a variety of ways. Electrophoresis with either ethidium bromide or

Table 1.6 Examples of sequence tagged sites characterized in barley. (After Chee *et al.*, 1993.)

Primer set	Primer sequences	Chromosomal location
aMSU21	5'GGTCTTTCATGTACCTACC3' 5'CGAGCTCCTGTCGAGG3'	2
β1-Hordein	5'CCACCATGAAGACCTTCCTC3' 5'TCGCAGGATCCTGTACAACG3'	5
A-Hordothionin	5'CTGGGGTTGGTTCTGG3'	5
Pst316	5'GCTTCAGAGAATGCATCTTG3' 5'CTGGTGAAGTACCTGATGAG3'	4
β-Glucanase	5'CCACCAAGCGTGGAGTC3' 5'GGGTGGCGTGGGGTG3'	5

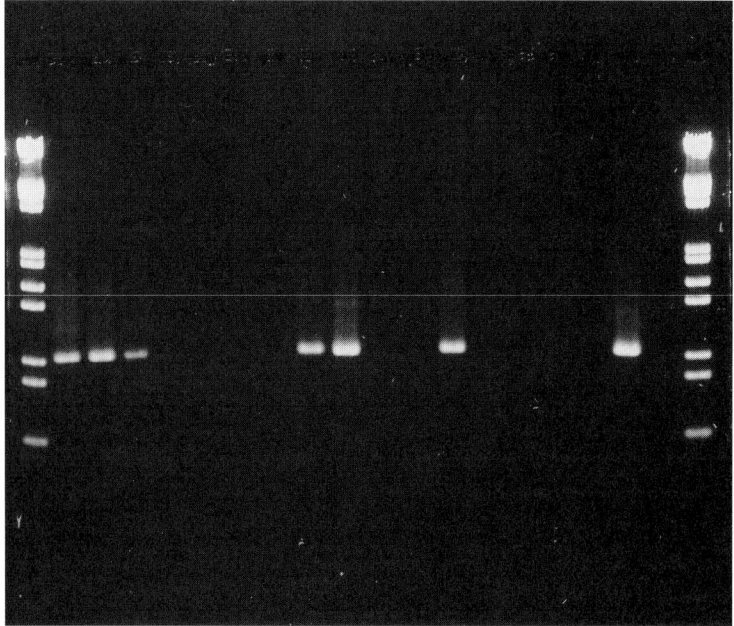

Figure 1.17 Sequence-characterized amplified region (SCAR) marker for a rust resistance gene from sunflower. A RAPD marker identified by bulked segregant analysis was cloned and sequenced. PCR primers designed from the sequence amplified a band only when the target gene was present. The marker tightly linked to the R1 rust resistance gene was used to analyse 17 genotypes, seven of which were indicated as having the gene.

silver staining is the general approach, but alternatives may have advantages in routine screening with molecular markers.

Electrophoresis may be avoided if only one PCR product is formed or if specific oligonucleotide probes are available for detection of the PCR product of interest. In cases where only one product can be amplified, ethidium bromide in the reaction tube can be used to detect positive amplifications under an ultra-violet light. Oligonucleotide probes may be designed to hybridize specifically to sequences within a PCR product of interest. This approach may be used in cases where many different PCR products result from the primers used. Probes labelled with digoxigenin (a steroid found in foxglove, *Digitalis purpurea*) may be detected using antibodies (Figure 1.18). A monoclonal antibody to digoxigenin may be conjugated to an enzyme capable of generating a coloured product. This allows the analysis of the PCR products in an ELISA (enzyme-linked immunosorbent assay) plate format. The automated equipment used for ELISA may be used

PCR–ELISA – designed to detect PCR amplification products in a
microtitre plate format.

1. PCR and labelling

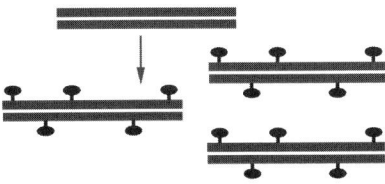

2. Denaturation and hybridization

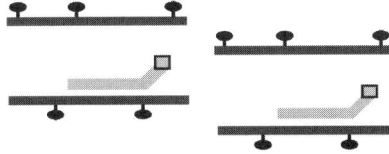

3. Capture and washing

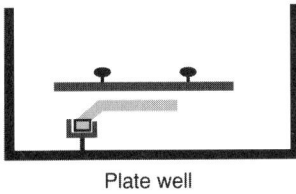

Plate well

4. Detection and reading

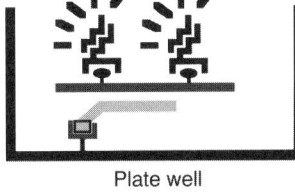

Plate well

(a)

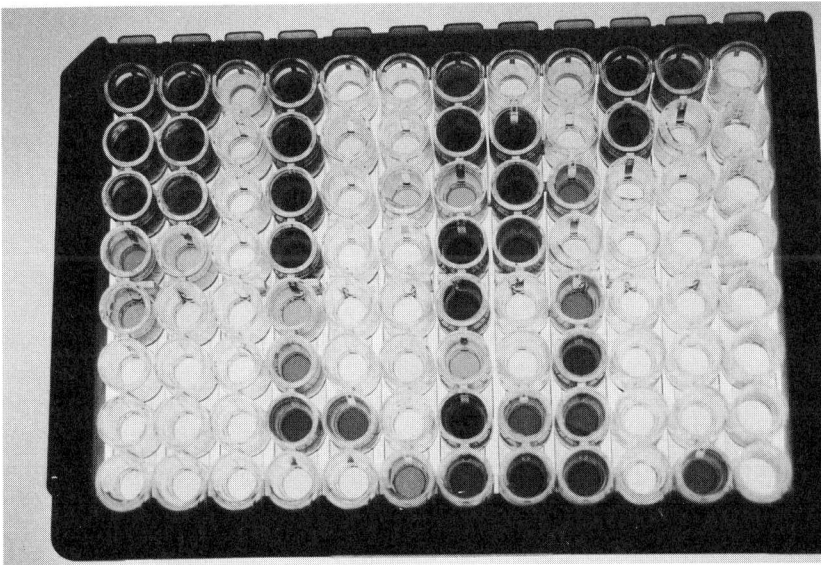

(b)

Figure 1.18 (a) and (b)

to detect the specific PCR product and to determine the amount
of product. This has the advantage of allowing the direct adoption of
the available technology (both hardware and software) for the capture,
analysis and storage of ELISA data for PCR product analysis.

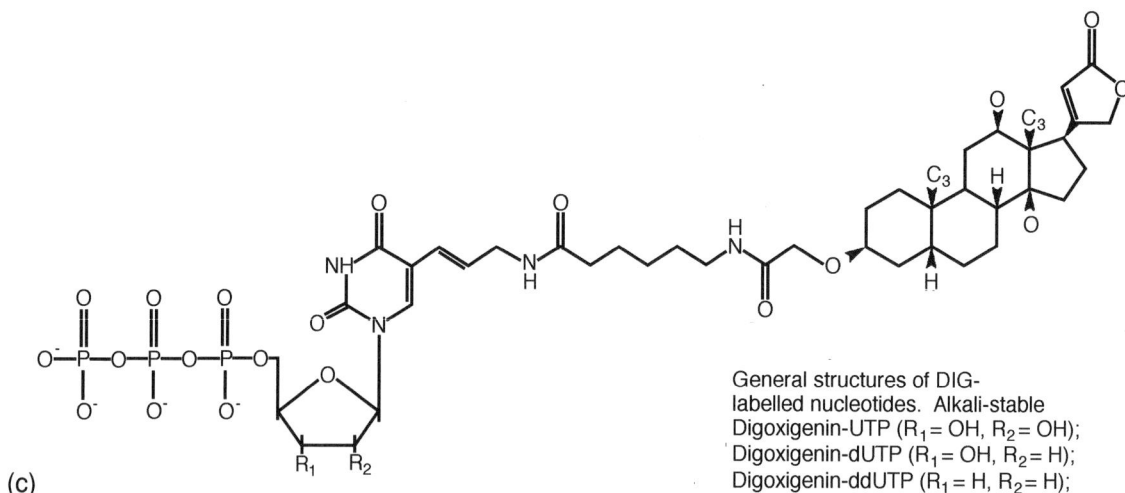

General structures of DIG-
labelled nucleotides. Alkali-stable
Digoxigenin-UTP (R_1 = OH, R_2 = OH);
Digoxigenin-dUTP (R_1 = OH, R_2 = H);
Digoxigenin-ddUTP (R_1 = H, R_2 = H);

(c)

Figure 1.18 Colorimetric detection of PCR products with digoxigenin (DIG)-
labelled probes. (a) Schematic representation of protocol. (b) Microtitre plate
analysis. (c) Structure of digoxigenin. PCR products generated from samples
of sugarcane juice were analysed by PCR using primers specific for the presence of *Clavibacter xyli,* the causal organism of ratoon stunting disease of
sugar cane. The PCR products were detected with a specific oligonucleotide
probe labelled with digoxigenin. Samples from diseased plants gave a colour
when the probe was detected using an antibody to digoxigenin (Boehringer-
Mannheim). The detection system is depicted schematically.

The exonuclease activity of DNA polymerases can be used to detect
specific PCR products directly (Holland *et al.*, 1991) (Figure 1.19). An
internal probe will be degraded by the enzyme during PCR. Blocking
of the 3' end of the probe with a phosphate will prevent PCR elongation. Detection of the degraded probe can be achieved by appropriate
labelling. A fluorescent label can be used as a reporter with a quencher
within the probe preventing fluorescence before degradation (Livak *et
al.*, 1995). This type of analysis provides simple DNA-based diagnosis
(Bassler *et al.*, 1995) in a single step and has potential for widespread
use in plant and plant pathogen identification. The advantages of this
system include the ability to multiplex (analyse for the presence of
several different sequences in the one tube by using fluorescent dyes
with different colours).

Probes that will fluoresce only when hybridized to the target molecule
have been devised (Tyagi and Kramer, 1996). These probes are complementary at the ends so as to form a hairpin stem when not hybridized.
The ends are linked to a fluorophore (5' end) and a quencher (3' end).
Disruption of the conformation by hybridization separates the fluorophore and the quencher, leading to fluorescence.

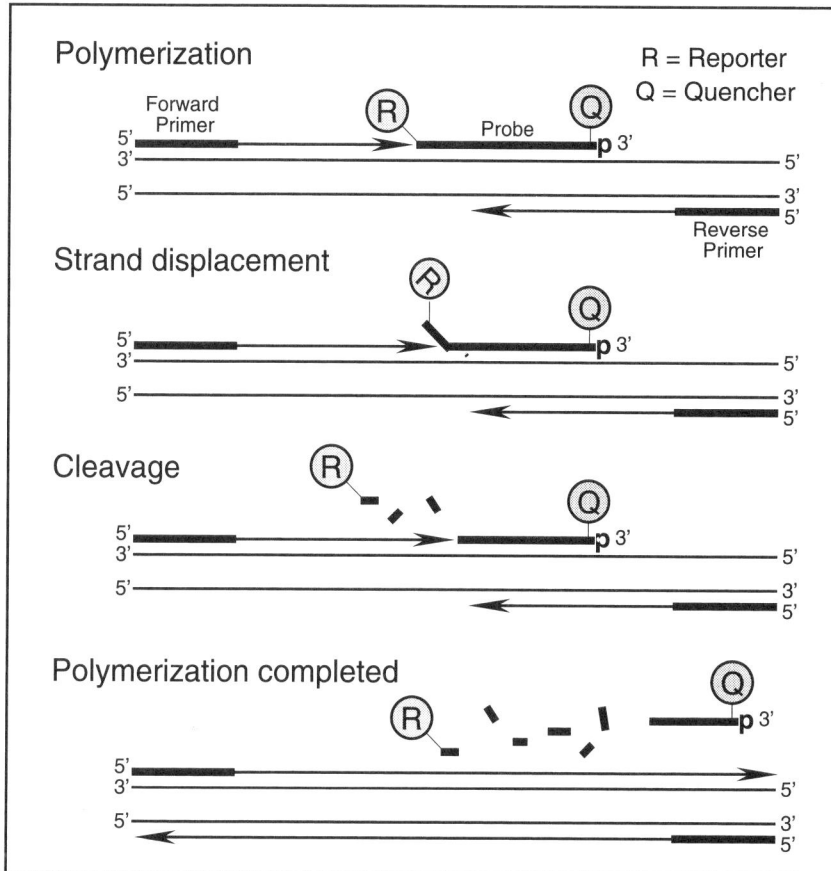

Figure 1.19 (a)

Direct colorimetric or fluorescent detection of PCR products is likely to be widely used for high-volume diagnostics because of the simplicity and speed of these techniques compared with electrophoresis.

1.1.13 ANALYSIS OF SEQUENCE VARIATION IN PCR PRODUCTS

The PCR products generated from different individuals may differ in sequence despite their length being similar or identical. Digestion of PCR products with restriction enzymes may reveal sequence poly-morphism in products of equal size. **Temperature gradient gel electrophoresis** (TGGE) may also be used to distinguish these products. **Heteroduplex analysis** involves the detection of sequence variations by

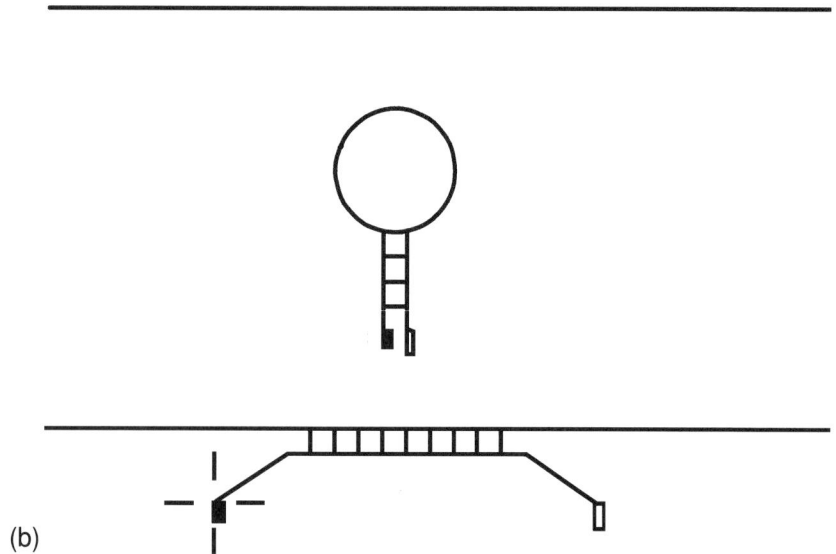

(b)

Figure 1.19 Detection of PCR products labelled with fluorogenic probes. (a) The endonuclease activity of TaqDNA polymerase is used to release a fluorescent reporter during PCR amplification in the TaqMan™ system (Perkin-Elmer). Fluorescence of the probe is quenched in the probe by a 3'-quencher dye. (b) Another approach involves the design of a probe with complementary sequences at the 5' and 3' ends. Hybridization to the target results in separation of the two ends revealing the fluorescence.

analysis following melting and reannealing in the presence of a reference sample. Products with a different sequence result in imperfectly matched hybridization to form heteroduplexes that can be easily separated by TGGE. This approach has been used to develop a system for the identification of barley varieties (Tsuchiya *et al.*, 1995). The principles of heteroduplex analysis are illustrated schematically in Figure 1.20.

Primers for analysis of chloroplast and mitochondrial DNA in plants have been described by Demesure *et al.* (1995) and these may be useful in heteroduplex analysis. The most suitable primers for widespread use are those based upon highly conserved sequences such as those in non-nuclear genomes and coding regions of the nuclear genome. Protocols using fluorescent labelling have been developed to allow efficient detection of point mutations following heteroduplex formation (Verpy *et al.*, 1994).

Sequence polymorphisms can also be detected by single-stranded conformational polymorphism (SSCP) analysis.

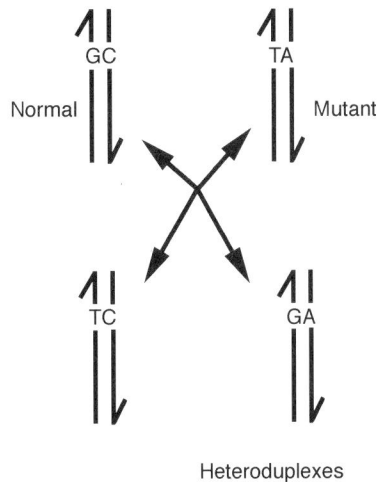

Normal   Mutant

GC   TA   TC   GA

Heteroduplexes

Figure 1.20 Heteroduplex analysis depicted schematically as described by Tsuchiya *et al.* (1994)

1.2.1 PRINCIPLES OF PLANT VARIETY PROTECTION

Plant breeders need to be able to identify commercial varieties to police the exploitation of their varieties and obtain a financial return from their investment in plant breeding. The granting of plant breeders' rights generally requires the establishment of the distinctness, uniformity and stability of the new variety. Proposed new varieties must be distinguished from all close relatives and known varieties. The uniformity of the variety must be shown to conform with generally accepted standards for varieties of the species concerned. Vegetatively, apomictic and fully and partially self-pollinated varieties are required to be more uniform than fully open-pollinated varieties. The variety must also be shown to be stable through at least two generations. Molecular techniques may be used to establish these properties for both obtaining and protecting plant breeders' rights (Morell *et al.*, 1995). UPOV (International Union for the Protection of New Varieties of Plants) is the international agency responsible for the protection of plant breeders' rights.

> **1.2 Molecular methods to protect plant breeders' rights**

1.2.2 TECHNIQUES TO VERIFY VARIETAL IDENTITY, PURITY AND STABILITY

Molecular marker techniques of all kinds may be used to define identity, purity and stability. The range of techniques available for immediate application varies, with major species offering a wide range of established molecular markers, and minor or less well-studied species requiring evaluation of potential marker systems. Arbitrary marker techniques such as RAPD analysis are attractive for the less well-known species because they can be applied without prior knowledge of gene sequences and in the absence of any genetic map or prior molecular studies. A useful method for the identification of plant varieties for the purpose of commercially protecting them should be based if possible on methods that can be reliably reproduced in any laboratory. Microsatellites may satisfy this requirement but their development is currently too expensive for most species, restricting their use to major crop species.

Molecular markers can be considered as additional characters or traits for the evaluation of genetic differences. The advantage of molecular markers is that DNA analysis is not influenced by the environment. Molecular data will identify differences in genotype, but these may not be used to determine issues such as the extent to which the new variety is substantially derived from earlier varieties. The concept of substantial derivation in the protection of plant varieties relates to traits of commercial value and not to genetic differences that do not alter the performance or value of the plant.

The distinction of plant varieties may be based upon differences in the variance of some characters rather than differences in their mean value. Molecular marker data may be analysed in the same way. The issues of adequate sampling in measurement of distinctness and uniformity are the same for molecular techniques as they are for the assessment of other characters such as morphological traits. A guide to the number of individuals that should be sampled to characterize a new variety is given in Table 1.7. The maximum number of off-types that is acceptable in a vegetatively propagated or fully self-pollinated variety may be defined as in Table 1.8. Double the number of off-types may be accepted in partially self-pollinated varieties. New open- or cross-

Table 1.7 Numbers of individuals to be assessed to characterize a new variety

Type of plant	No. of plants		
	Minimum	Maximum	Replicates
Cloned trees	5	10	5*
Cloned horticultural species/vegetables	30	60	2
Self-pollinated species	30	30	3
Open pollinated species	100		2+

*One plant per replicate for cloned trees.

Table 1.8 Numbers of off-types acceptable in new varieties of vegetatively or fully self-pollinated varieties

Sample size	Maximum no. of off-types*
5	0
6–35	1
36–82	2
83–137	3

*The number is doubled for partially self-pollinated varieties.

pollinated varieties should exhibit variation that does not exceed greatly that found in existing varieties.

The need to assess genetic stability is related to the method of propagation of the plant. Vegetatively propagated species should show very little variation between generations. Tissue culture may introduce some genetic changes and sexual reproduction can be the basis of very large differences. Again, the application of molecular techniques to the assessment of genetic stability does not differ in principle from the assessment of other non-molecular traits.

The main requirements of any molecular test to be used in plant varietal protection are: (i) that the test be repeatable, both within and between laboratories; (ii) that they may be objectively assessed or scored; and (iii) that they provide satisfactory discrimination between varieties (Smith and Chin, 1992).

1.2.3 EXAMPLES

(a) Soybean variety analysis using SSR

Three microsatellite loci with $(AT)_n$ and four loci with $(ATT)_n$ were applied to 96 soybean genotypes by Rongwen *et al.* (1995). Soybean varieties are increasingly difficult to distinguish as they often arise from a narrow range of elite parents with very small genetic differences. Only two of the varieties had the same profiles at the 11 to 26 alleles found at the seven loci. The two varieties that were not distinguished had very similar pedigrees. The similarities of the 96 varieties varied from 0.71 to 0.95 with a mean of 0.87. This provides a useful system for the identification of soybean varieties and the protection of the commercial interests of soybean breeders.

Figure 1.21 Navy beans: molecular methods help to distinguish varieties with very similar morphological features.

Navy bean variety analysis using RAPD

Three new navy bean (*Phaseolus vulgaris* L.) varieties were evaluated using the RAPD method by Graham *et al.* (1994). These varieties are very difficult to distinguish using morphological traits, making them an ideal candidate for the application of molecular marker methods (Figure 1.21). The breeding history is an important consideration in approaching this type of analysis. The varieties were produced by early generation bulk breeding followed by single plant selection in the F_4. At F_9, single plants were examined and only phenotypically similar material was bulked. The expected heterogeneity was low because of this approach, suggesting that relatively small samples (relatively few individuals) might be suitable for use in variety distinction. Sirus was from a cross between Campbell 18 and BAC 134, Rainbird was from Rufus × Actolac and Spearfelt was from Campbell-11 × CH33-8D. The parental varieties, Campbell-18, BAC 134, Rufus, Actolac, Campbell-

11 and CH33-8D were also included in the study. Two samples of Actolac were analysed as a check on variations that might be found when analysing two samples of the same variety. DNA was extracted from bulked material from 10 seedlings of each variety. The 12 samples were analysed using 60 primers. This produced 296 markers that could be scored. The average similarity was about 85%. Rainbird and Sirus were very similar (96% similarity). The two Actolac samples differed by less than 2%. This indicates that differences of this magnitude could not be used reliably to distinguish varieties.

Uniformity was assessed by analysing 10 individuals of each variety. The markers were in most cases consistent for at least nine of the 10 plants. More than one off-type in 10 plants would make the marker unreliable for variety distinction.

All 11 navy bean varieties could be readily distinguished using just two of the primers.

(c) Other examples using RAPD

Many other examples of application of RAPD to identification of plants for the protection of plant varieties have been reported. Examples include application to apple varieties (Tancred *et al.*, 1994) and rose varieties (Torres *et al.*, 1993).

1.3.1 NEED FOR PATHOGEN IDENTIFICATION

Poor plant performance may be due to many factors, adverse environmental conditions, inadequate nutrition or pests and diseases. Figure 1.22 shows a barley plant infected with spot blotch (*Bipolaris sorokiniana*). The diagnosis of the cause of such diseases can depend upon the reliable identification of specific infective agents. The pathogen may be a virus, bacterium, fungus, nematode, or a combination of these. The symptoms of these diseases and of environmental factors may not always be readily distinguished. Molecular methods provide generic techniques for distinction between these possibilities and the identification of specific species or genotypes of the pathogens.

1.3.2 TECHNIQUES FOR PATHOGEN IDENTIFICATION

The full range of molecular methods are possibilities for use in pathogen identification. Techniques such as ribotyping (analysis of ribosomal RNA genes), analysis of plasmid DNA content, RAPD–PCR (or related techniques) and pulsed field gel electrophoresis following digestion with rare cutting endonucleases may be used to identify

> ### 1.3 Identification of plant pathogens

Figure 1.22 Diseased plant: molecular methods can be used to identify the presence of a specific pathogenic organism in a diseased plant sample.

microorganisms. The analysis of pure cultures of these organisms is a starting point in the development of a diagnostic method, but ultimately a test that can be applied directly to diseased plants may be necessary. Application to a diseased plant requires that the method be based upon sequences that distinguish the pathogen from the host plant. Root diseases may require protocols for the analysis of soil samples. Greater sensitivity can be achieved by basing tests upon multi-copy genes and by using appropriate labelling of probes or primers.

1.3.3 EXAMPLES OF MOLECULAR METHODS FOR PLANT PATHOGEN DIAGNOSIS

(a) Specific probes

RFLP analysis has been used to distinguish *Sclerotinia* species (Kohn *et al.*, 1988). Both nuclear and mitochondrial genes were useful probes for these distinctions. A ribosomal RNA gene probe from *Neurospora crassa* distinguished all but two of seven species of *Sclerotinia*. Two species showed some intra-specific variation with this probe. Probes derived from the ribosomal RNA genes of two other species, *Schizophyllum commune* and *Armillaria mella*, hybridized to exactly the same fragments. A high level of polymorphism was detected by a probe from the mitochondrial 24S ribosomal RNA from *Neurospora crassa*.

Pithium causes root disease in a wide variety of plant species. A repeat sequence from a genomic library of *Pithium irregulare* has been used to identify this species in soils (Matthew *et al.*, 1995).

Detection and distinction of plant viruses may be easily achieved using specific probes. The peanut stripe virus and the peanut mottle virus may be detected and distinguished in samples of peanut (*Arachis hypogaea*) using probes from the 3' terminal of the viruses (Dietzgen *et al.*, 1994). Digoxigenin-labelled cRNA probes of 1400–1700 nucleotides were able to provide a high level of specificity and sensitivity for the detection of these diseases.

(b) PCR-based methods

PCR has been widely applied to the diagnosis of plant pathogens (Henson and French, 1993; Honeycutt and McClelland, 1996). The advantages of PCR include specificity and sensitivity. The ultimate PCR method could distinguish a single base difference and be sensitive enough to detect a single molecule or individual. The test must be totally reliable in always detecting the pathogen when present but never giving a positive result in the absence of the pathogen. The inclusion of positive controls is an important approach to reducing the incidence of false negatives due to failures in the reagents. False positives may be more difficult to eliminate, but careful sample handling and negative controls may reduce their incidence.

For example, PCR methods have been developed for the detection and distinction of *Gaeumannomyces graminis* var. *tritici* and var. *avenae*, pathogenic on wheat and oats respectively, causing the disease take-all (Ward, 1995). These methods include an internal standard that is important in insuring against false negatives. PCR reactions may fail if the sample to be tested contains PCR inhibitors. Unpurified DNA, especially from samples of diseased plant tissues such as roots (potentially contaminated with soil), may cause these problems. An internal control PCR can detect the presence of these inhibitors and avoids errors in incorrectly diagnosing the absence of the pathogen in cases of PCR failure.

(c) Arbitrary primers

PCR with arbitrary primers has the advantage of being applicable in the absence of any sequence information, but may require culture of the organism because of potential interference from the host plant DNA.

(d) RAPD

RAPD markers have been widely used to distinguish plant pathogens. Isolates of *Colletotrichum graminicola*, the causal organism of anthracnose of sorghum, have been characterized in this way (Guthrie *et al.*, 1992). Three primers were used to generate distinguishing profiles for isolates collected from sorghum growing in Africa, the United States, India, Puerto Rico and from Johnson grass in the USA. Image analysis was used with a threshold value for band intensity below which the band was rejected. The gel image was converted to a barcode by further processing. Amplification on separate days was shown to give a consistent profile after image processing. Pooled data from three primers was shown to be more reliable than data from a single primer alone. This provided a protocol with potential application in disease quarantine.

(e) SCAR

Sequencing of markers generated with arbitrary primers can be used to develop more robust PCR methods for pathogen diagnosis. These SCAR markers have been used, for example, to characterize powdery mildew (*Erysiphe graminis* f. sp. *hordei*) isolates from barley (McDermott *et al.*, 1994). SCAR markers were found to have special value in the detection of mixtures of isolates.

(f) Ribosomal gene analysis

RFLP of ITS amplified by PCR has been used to distinguish *Pythium* species (Chen, 1992). The ITS varied in length and digestion of the amplified ITS with restriction enzymes allowed further discrimination. Differences in ITS sequence have been used to develop primers to specifically diagnose *Verticillium* species causing wilting diseases of potato (Moukhamedov *et al.*, 1994). PCR of tRNA intergenic spacers has been used as a diagnostic test for *Xanthomonas albilineans*, a pathogen of sugarcane (Honeycutt *et al.*, 1995)

(g) PCR analysis of plant viruses

Plant viruses may be detected using specific PCR. Reverse transcriptase–polymerase chain reaction (RT–PCR) has been used for RNA viruses. Immunocapture has been used to separate the virus from other contaminating substances in the plant. A technique for immobilization of the virus on polystyrene or polypropylene surfaces followed by RT–PCR was used by Rowhani *et al.* (1995) to detect viruses in plant material. Protocols for the direct detection of both DNA and RNA viruses have been developed based upon PCR of a simple extract of the plant tissue (Thomson and Dietzgen, 1995).

1.4.1 INTRODUCTION

> **1.4 Commercial, industrial, forensic and other plant identification**

Plant identification may be important in ensuring authenticity of genotypes in commercial trading, especially for horticultural or ornamental species. The processing of plants to produce food or industrial products may make identification by conventional techniques difficult, if not impossible. DNA analysis has potential for application in a wide variety of industrial situations. Food processing may require a knowledge of the species or varietal composition or raw materials or partial or fully processed ingredients derived from plants. This information may be of value in process control, quality control and in ensuring functional and nutritional properties of the food. Labelling of food products may be verified by DNA analysis.

1.4.2 IDENTIFICATION OF HORTICULTURAL PLANTS

The identity of horticultural genotypes may be of considerable commercial importance (Figure 1.23). Planting an orchard of the incorrect genotype could be a costly mistake that may not be revealed for several years. The confusion of the identity of an apple variety in commercial

Figure 1.23 Fruit varieties: molecular methods can be used to ensure that correct varieties are propagated and produced.

production has been easily resolved using RAPD markers (Tancred *et al.*, 1994). The first sign of this problem was the apparent susceptibility of apples to diseases to which the variety was considered resistant. This prompted a genotype test using molecular markers, immediately indicating that the genotype in production was different and suggesting a confusion of budstocks during propagation. Molecular methods have the potential to be used in the verification of genotype identity in commercial propagation.

1.4.3 SPECIES COMPOSITION OF FOOD

Analysis of the species composition of foods has commercial application in verifying the composition claimed on the label and in the commercial assessment of ingredients used in competitors' products. The species composition of feeds may be commercially useful information, especially if a competitor has discovered a less expensive ingredient. Contamination of foods with species with toxic or anti-nutritional species, or the presence of species that cause allergic response may be important to specific consumers.

Molecular analysis designed to detect species-specific sequences has the potential reliably to detect the presence of very low levels of specific

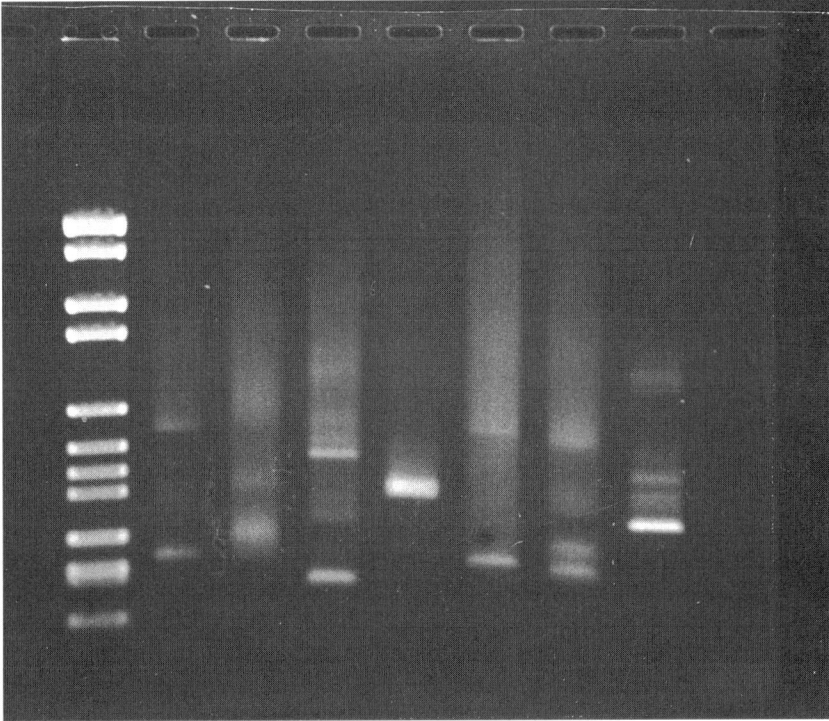

Figure 1.24 Specific detection of cereal species by PCR. The first lane is a size marker. The remaining lanes are the result of amplification of 5S ribosomal RNA spacers from different cereal species; left to right, wheat, barley, oat, rye, rice, sorghum and maize.

species in foods. Ribosomal gene analysis, especially by PCR of 5S RNA gene spacers, is a useful option for such cases (Figure 1.24).

A general approach is to use a species-specific probe to identify the presence of each species. Many specific probes are available. Example of food species for which specific probes have been reported include, kiwifruit (*Actinidia deliciosa*; Crowhurst and Gardner, 1991), tomato (*Lycopersicon esculentum*; Schweizer *et al.*, 1988) and the major cereals (Zhao *et al.*, 1989).

Specific primers for the detection of plant species may be designed by sequencing the spacers between the 5S ribosomal RNA genes (these spacers may be amplified using universal primers as shown in Figure 1.25). PCR with the species-specific primer is depicted schematically in Figure 1.26. Results of this type of analysis are shown in Figure 1.27 A single PCR band indicates the presence of the target species. This has been proposed as a general method for plant identification in samples of unknown or mixed species (Ko and Henry, 1996). The

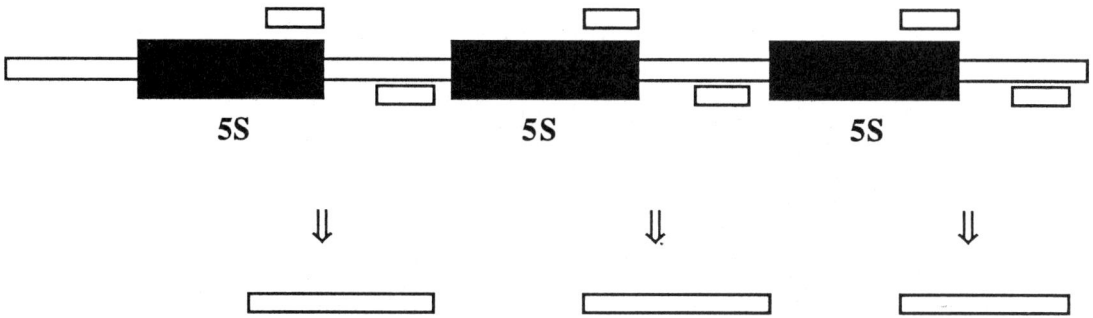

Figure 1.25 Design of a species-specific PCR detection based upon 5S ribosomal gene spacers. A primer from the conserved region of the 5S ribosomal RNA gene is paired with a species-specific primer from the spacer region. The unique species-specific primer sequence is obtained by sequencing the main spacer band amplified as depicted in Figure 1.10(b) and shown in Figure 1.24.

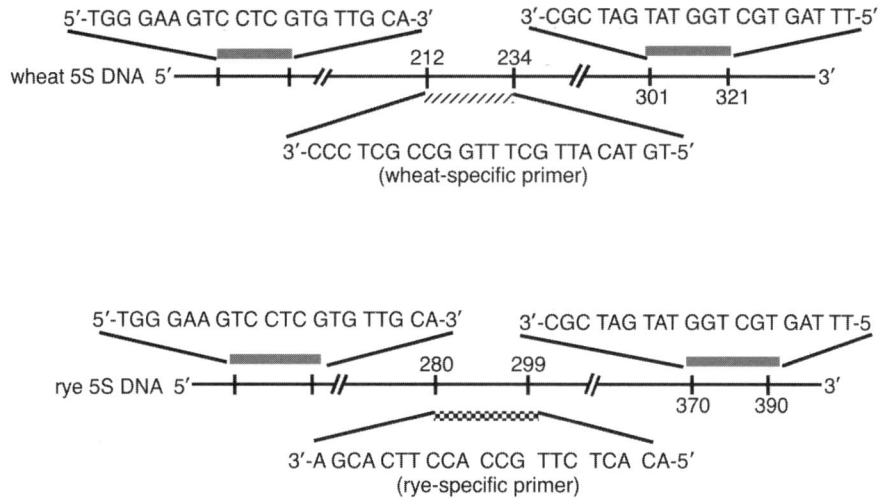

5′-TGG GAA GTC CTC GTG TTG CA-3′   3′-CGC TAG TAT GGT CGT GAT TT-5′

212   234

wheat 5S DNA 5′—————//————//————3′

301   321

3′-CCC TCG CCG GTT TCG TTA CAT GT-5′
(wheat-specific primer)

5′-TGG GAA GTC CTC GTG TTG CA-3′   3′-CGC TAG TAT GGT CGT GAT TT-5

280   299

rye 5S DNA 5′—————//————//————3′

370   390

3′-A GCA CTT CCA CCG TTC TCA CA-5′
(rye-specific primer)

Figure 1.26 Primers designed to detect specifically the presence of wheat and rye.

lack of amplification in the absence of the species indicates that, instead of electrophoresis, a method for detection of amplified DNA (e.g. ethidium bromide) could be used to establish the presence of the target species. Other colorimetric detection methods could also be used to determine the presence of the diagnostic PCR product and simplify the test format.

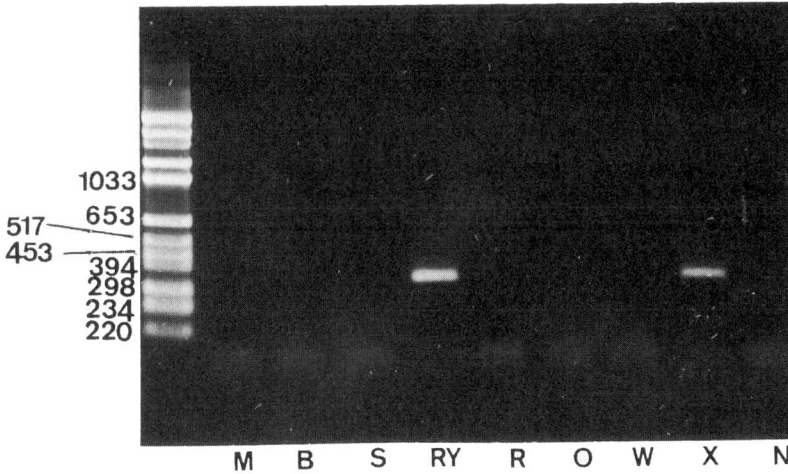

Figure 1.27 Species-specific PCR amplification using primers to 5S ribosomal gene spacers. Results for a rye-specific primer are shown. M, maize; B, barley; S, sorghum; RY, rye; R, rice; O, oat; W, wheat; X, mixture of all seven species, N, negative control (no template).

1.4.4 VARIETAL IDENTITY AND PURITY IN GRAIN PROCESSING

The identification of variety may also be important in food processing. The cereal species represent a large part of all human food. More than 500 million tonnes each of wheat, rice and maize are produced annually and consumed directly or as a wide range of processed foods such as breads, noodles and pasta. Barley is the fourth-ranked cereal in production and represents the major raw material for the production of beer and other beverages. The distinction of cereal varieties with different processing properties is an example where the identification of the variety by morphological methods may not be possible. This is important in the storage and transport of grain so as to ensure the reliable segregation of grain lots with different properties and market values (Figure 1.28). Wheat varieties with similar or identical appearance may have very different suitabilities for use in the production of specific end-products such as noodles. The difference in end-use quality will in many cases result in large differences in the value of the grain, creating an incentive to substitute or blend varieties of lower value. Similar problems often arise in relation to the distinction of high-value malting barley varieties (used for beer production) from lower-value feed varieties (used for animal feed) with identical appearance. PCR with alpha-amylase primers has allowed identification of barley varieties and distinction of closely related feed and malting varieties (Ko

Figure 1.28 Grain silos. The storage, transport and marketing of grain can be aided by accurate testing of varietal identity. The processing characteristics of specific genotypes can used to advantage in the production of specific end-products (e.g. bread, noodles, beer).

and Henry, 1994b). The detection of common wheat contaminants in durum wheat is important in some markets. Food labelling laws in many countries require pasta products (Figure 1.29) to be made only from durum wheat. DNA analysis offers an ideal solution to the establishment of identity and purity in these circumstances. Methods suitable for analysis of single grains by PCR have been developed (Benito *et al.*, 1993). RAPD analysis or if available SSR approaches have been used in these situations.

1.4.5 IDENTITY OF GRAPE VARIETIES

Wine production is based upon the many varieties of grape (*Vitus vinifera*) with the identity of the variety being of considerable commercial significance in high-quality (value) wines (Figure 1.30). Grape varieties have been allocated more than 24 000 names but as few as 5000 genotypes may exist. Grapes have spread throughout the world by vegetative propagation and been given local names. Microsatellites have been used to distinguish varieties (Thomas and Scott, 1993) and will eventually allow the number of grape variety names to be reduced and confusion about the identity of varieties and differences between varieties grown in different wine producing areas of the world to be resolved. An international database of variety microsatellite profiles has been produced and validated in the distinction of phenotypically similar varieties (Thomas *et al.*, 1994).

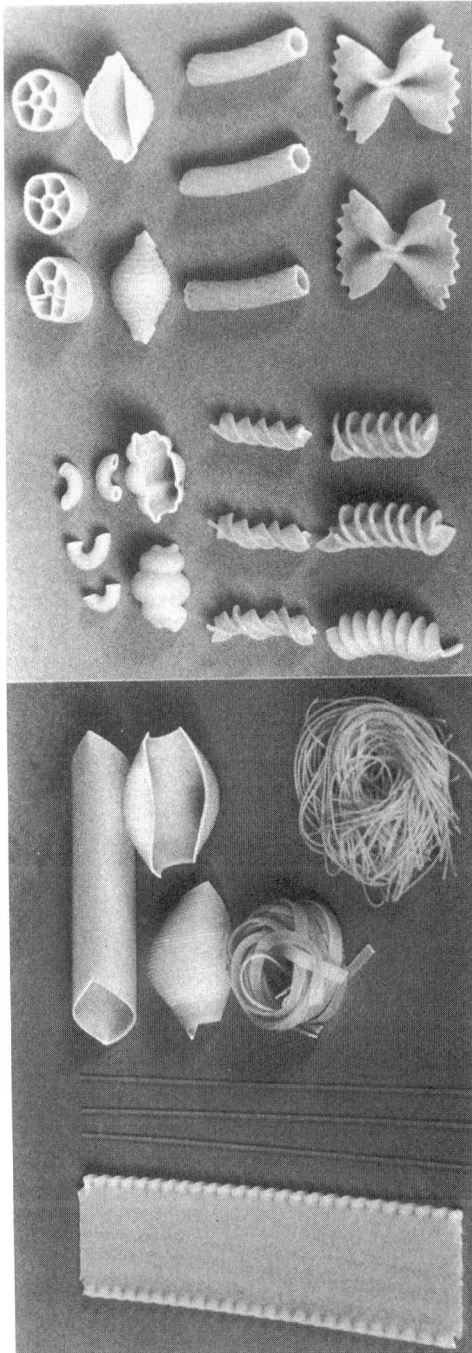

Figure 1.29 Pasta. The purity of durum wheat used for pasta production may be regulated. PCR-based tests have the potential to detect adulteration of flour with common (hexaploid) wheat genotypes.

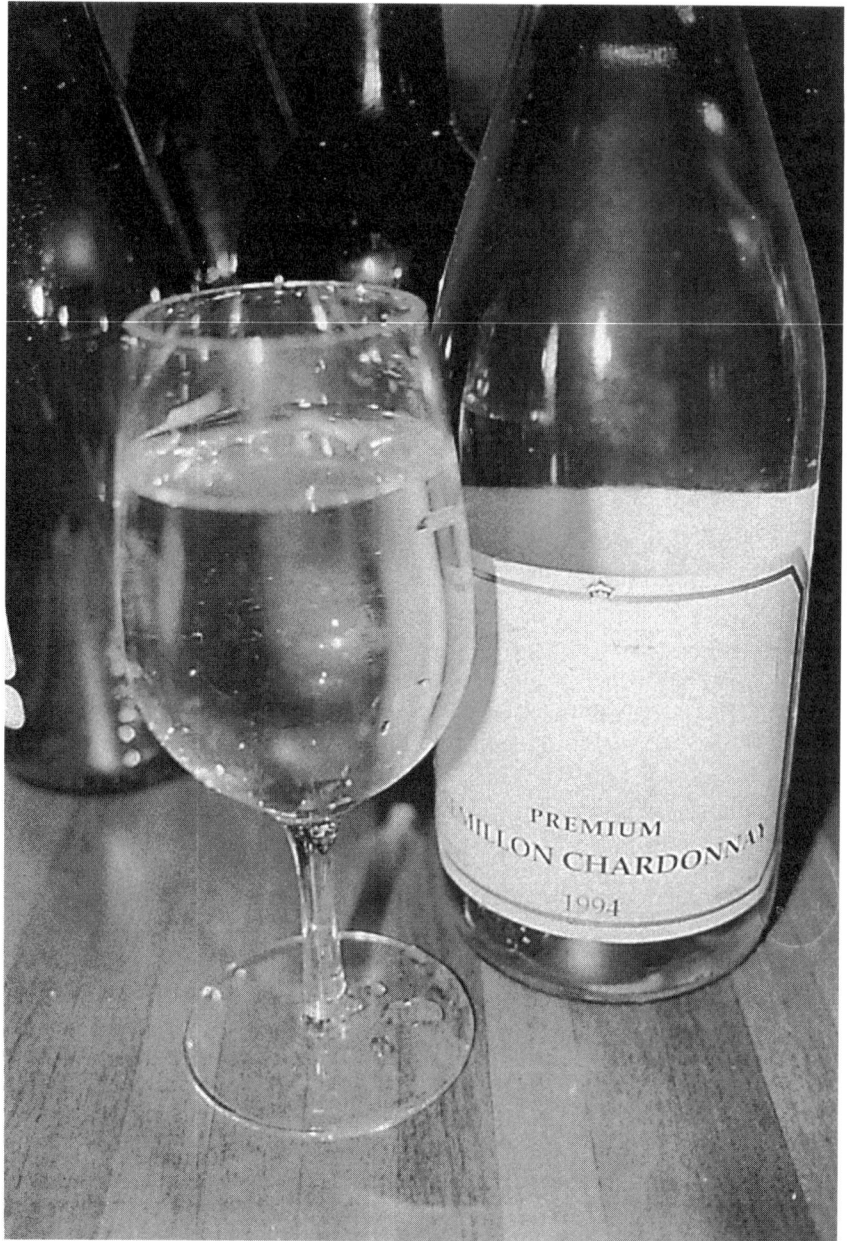

Figure 1.30 The identity of grape varieties may be important to the wine maker.

1.4.6 DETECTING OFF-TYPES IN PLANT PROPAGATION BY TISSUE CULTURE

Plant tissue culture is used to propagate many ornamental and crop species (Figure 1.31). Somaclonal variation in some species results in inferior off-types with little commercial value. Somaclonal variation may result from changes in nuclear, mitochondrial or chloroplast genomes. Molecular methods may be used to screen plants derived from tissue culture to eliminate individuals with these defects. Banana tissue culture frequently generates dwarf off-types with no commercial value (Figure 1.32). This problem has restricted the widespread adoption of tissue culture for the production of planting material, despite the advantages in disease control offered by this approach. A RAPD marker has been identified for this common off-type (Damasco *et al.*, 1996). A routine PCR method was developed following sequencing of the RAPD band distinguishing the dwarf plants. Similar approaches are likely to be of value in quality control of other plants propagated in tissue culture. Somaclonal variation in sugarcane (Taylor *et al.*, 1995) and *Populus deltoides* (Rani *et al.*, 1995) has been detected using RAPD techniques. Molecular techniques can be used to define culture conditions, thus minimizing genetic variation. These techniques are important in the genetic engineering of plant varieties with one or a few new genes, but retaining the genetic background of an elite variety. Molecular approaches may be used to monitor genetic integrity during culture and transformation and to develop protocols to eliminate or minimize genetic change. The full range of molecular techniques described in this chapter could be used to monitor somaclonal variation. The choice of technique will depend upon factors such as the level and type of genetic variation to be detected (Henry, 1997). Somaclonal variation may involve changes ranging from major chromosomal rearrangements to single point mutations. The nuclear, chloroplast or mitochondrial genome may undergo somaclonal variation. The extent of somaclonal variation increases with time in culture. Molecular markers such as RFLP (Muller *et al.*, 1990) and RAPD (Taylor *et al.*, 1995) analysis have been used to follow the development of somaclonal variation. The increasing body of knowledge on plant genome structure and the sequences of plant genes will provide a background database against which somaclonal variation can be measured.

1.4.7 OTHER APPLICATIONS

Forensic applications of plant DNA analysis may arise when plant material at a crime scene can be tested against plant material found,

(a)

(b)

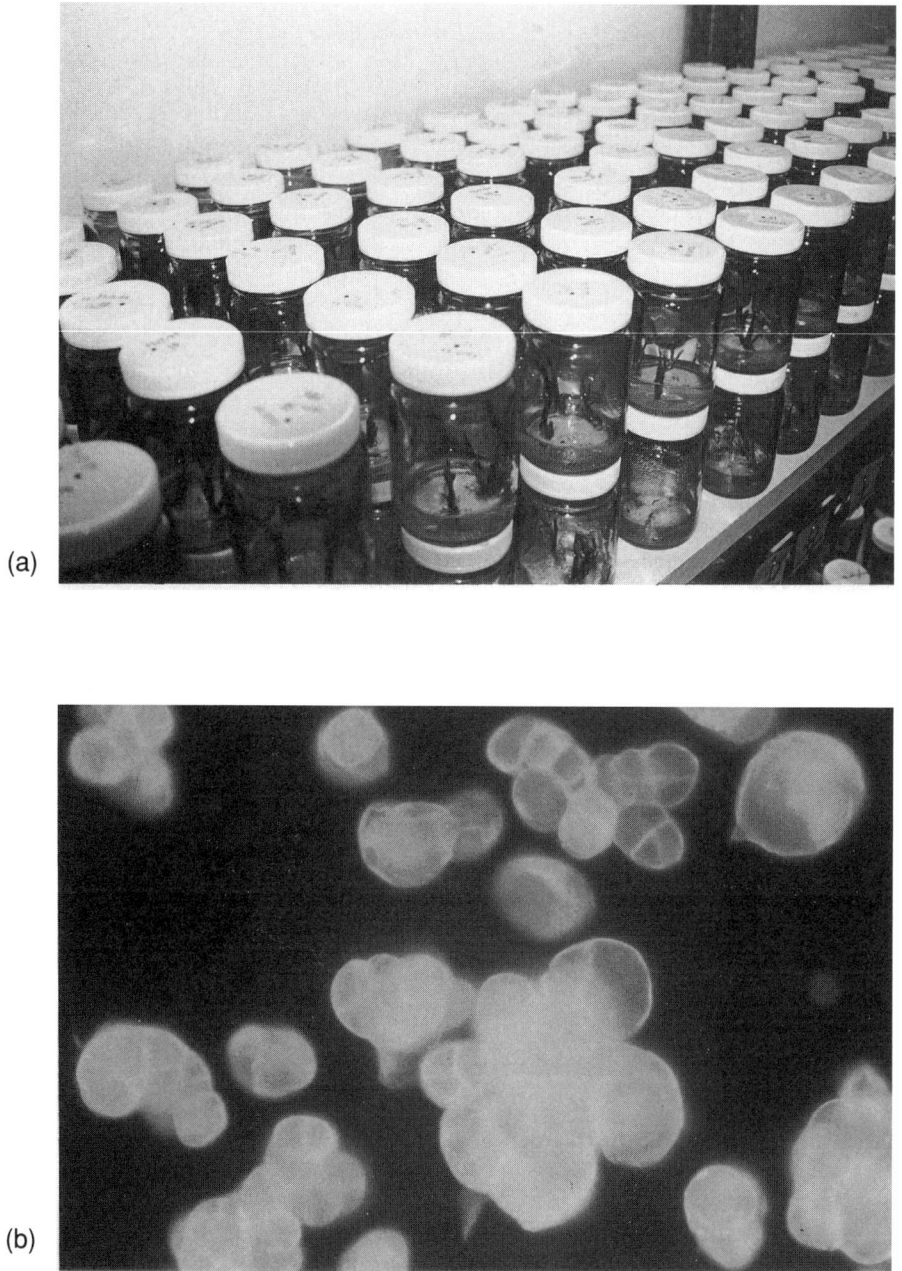

Figure 1.31 Cell and tissue culture of plants. Molecular markers may be used to monitor somaclonal variation and avoid the propagation of sports. (a) Propagation of banana by tissue culture. (b) Suspension cultured ryegrass (*Lolium multiflorum*) cells. This type of culture can result in rapid genetic change.

Figure 1.32 Dwarf somaclonal variants of bananas. Molecular methods may be used to distinguish off-types during culture. (a) Normal and (b) dwarf variants of the New Guinea Cavendish banana cultivar.

for example, on the clothing of a suspect. This may involve identification of the species, variety or individual plant by DNA profiling.

Positive distinction between weeds and closely related species may be important in devising weed control strategies.

Determination of the stomach content of animals may prove useful in studies of the composition of the diet of grazing animals and in establishing the presence of poisonous species. Analysis of the diet of grazing animals with a range of plant species available for feeding may indicate the preferences of the animal. These studies may be used to explain differences in animal growth rates and deduce feed values contributed by different plant species. Toxic plant species may be identified at autopsy.

Plant identification may become important in plant quarantine. The identity of samples being transported may require verification to avoid the introduction of unwanted weeds or plant diseases. Contamination of cargoes such as fertilizers with seeds such as grain from previous cargoes carried in the same hold can result in large financial losses and require careful identification of the contaminating seed.

The widespread application of plant variety identification in plant breeding will be covered in Chapter 3.

KEY TERMS

cDNA
DNA amplification fingerprinting (DAF)
Genomic DNA
Heteroduplex analysis
Inter geneic spacer (IGS)
Internal transcribed spacer (ITS)
Intron splice junction (ISJ)
Microsatellites
Minisatellites
Polymerase chain reaction (PCR)
Polymorphism
Primer
Probe
Random amplified polymorphic DNA (RAPD)
Restriction fragment length polymorphism (RFLP)
Sequence characterized amplified region (SCAR)
Sequence tagged site (STS)
Short sequence repeat (SSR)
Temperature gradient gel electrophoresis (TGGE)
Variable number of tandem repeats (VNTR)

EXAMPLES OF WORKED QUESTIONS

1. Define the following acronyms: RFLP, SCAR and SSR.

   RFLP, Restriction fragment length polymorphism; SCAR, Sequence characterized amplified region; SSR, Short sequence repeat.

2. Describe how you would determine the purity of a seed lot by RAPD analysis.

   RAPD–PCR may be applied to the analysis of DNA from individual seeds and the products compared with those obtained from samples of known varieties. The assessment should include enough primers to allow distinction of the variety from all possible contaminating varieties and to evaluate adequately any variation within the variety.

3. Name three molecular marker methods that could be used to identify a plant pathogen in a diseased plant sample by first culturing the pathogenic organism. Which of these methods could be applied directly to the analysis of a diseased plant sample?

   RFLP, RAPD and SCAR or STS. The RFLP and SCAR methods would not necessarily require the separation of the plant and pathogen DNA. RAPD analysis could result in amplification of fragments from the plant DNA that might interfere with the detection and identification of the pathogen.

4. How many plants would you sample to characterize a new apple variety with a molecular marker in order to protect the plant breeder's rights?

   Five plants as a minimum and preferably ten plants.

5. List three methods available for the labelling of a DNA probe

   Radiolabelling with ^{32}P, labelling with digoxigenin and fluorescent labelling.

6. Describe possible applications of molecular methods in grain trading.

   Molecular methods could be used to check the varietal identity of grain lots and thus determine their suitability for specific end-uses and to detect the presence of admixtures of grain of inferior varieties.

Questions

1. Define the following acronyms: RAPD, AFLP and STS.

2. Describe how you would determine the plant species that had been used to manufacture a confectionery product.

3. Name three molecular methods that could be applied to the identification of a plant variety.

4. How many off-types would be acceptable in five seeds of a new variety of wheat evaluated with a set of microsatellite markers?

5. List three PCR-based molecular marker methods.

6. Describe possible uses of molecular methods for grape analysis in the wine industry.

References

Bassam, B.J. and Bentley, S. (1994) DNA fingerprinting using arbitrary primer technology (APT): a tool or a torment? *Australasian Biotechnology*, 4, 232–6.

Bassler, H.A., Flood, S.J.A., Livak, K.J., Marmaro, J., Knorr, R. and Batt, C.A. (1995) Use of a fluorogenic probe in a PCR-based assay for the detection of *Listeria monocytogenes*. *Applied and Environmental Microbiology*, 61, 3724–8.

Benito, C., Figueiras, A.M., Zaragoza, C., Gallego, F.J. and de la Pena, A. (1993) Rapid identification of Triticeae genotypes from single seeds using the polymerase chain reaction. *Plant Molecular Biology*, 21, 181–3.

Brunel, D. (1994) Denaturing gradient gel electrophoresis (DGGE) and direct sequencing of PCR amplified genomic DNA: rapid and reliable identification of *Helianthus annuus* L. cultivars. *Seed Science and Technology*, 22, 185–94.

Caetano-Anoles, G., Bassam, B.J. and Gresshoff, P.M. (1991) DNA amplification fingerprinting using very short arbitrary oligonucleotide primers. *BioTechnology*, 9, 553–7.

Chee, P.W., Pederson, L., Matejowski, A., Kanazin, V. and Blake, T. (1993) Development of polymerase chain reaction for barley genome analysis. *American Society of Brewing Chemists Journal*, 51, 93–6.

Chen, W. (1992) Restriction fragment length polymorphisms in enzymatically amplified ribosomal DNAs of three heterothallic *Pythium* species. *Phytopathology*, 82, 1467–72.

Crowhurst, R.N. and Gardner, R.C. (1991) A genome-specific repeated sequence from kiwifruit (*Actinidia deliciosa* var. *deliciosa*). *Theoretical and Applied Genetics*, 81, 71–8.

Damasco, O., Graham, G.C., Henry, R.J., Adkins, S.W., Smith, M.K. and Godwin, I.D. (1996) Random amplified polymorphic DNA (RAPD) detection of dwarf off-types in micropropagated Cavendish (Musa spp. AAA) bananas. *Plant Cell Reports*, 16, 118–22.

Demesure, B., Sodi, N. and Petit, R.J. (1995) A set of universal primers for amplification of polymorphic non-coding regions of mitochondrial and chloroplast DNA in plants. *Molecular Ecology*, **4**, 129–31.

Denda, T., Kosuge, K., Watanabe, K., Ito, M., Suzuki, Y., Short, P.S. and Yahara, T. (1995) Intron length variation of the *Adh* gene in *Brachyschome* (Asteraceae). *Plant Molecular Biology*, **28**, 1067–73.

Devos, K.M., Bryan, G.J., Collins, A.J., Stephenson, P. and Dale, M.D. (1995) Application of two microsatellite sequences in wheat storage proteins as molecular markers. *Theoretical and Applied Genetics*, **90**, 247–52.

Dietzgen, R.G., Zeyong, X. and Techeney, P.-Y. (1994) Digoxigenin-labeled cRNA probes for the detection of two polyviruses infecting peanuts (*Arachis hypogaea*). *Plant Disease*, **78**, 708–11.

D'Ovidio, R. (1993) Single-seed PCR of LMW glutenin genes to distinguish between durum wheat cultivars with good and poor technological properties. *Plant Molecular Biology*, **22**, 1173–6.

D'Ovidio, R., Tanzarella, O.A., Masci, S., Lafiandra, D. and Porceddu, E. (1992) RFLP and PCR analyses at *Gli-1*, *Gli-2*, *Glu-1* and *Glu-3* loci in cultivated and wild wheats. *Hereditas*, **116**, 79–85.

Gepts, P., Stockton, T. and Sonnante, G. (1992) Use of hypervariable markers in genetic diversity studies, in *Applications of RAPD Technology to Plant Breeding*, Crop Science Society of America, American Society for Horticultural Science, American Genetic Association, Minneapolis, pp. 41–5.

Graham, G.C., Henry, R.J. and Redden, R.J. (1994) Identification of navy bean varieties using random amplification of polymorphic DNA. *Australian Journal of Experimental Agriculture*, **34**, 1173–6.

Guidet, F., Rogowsky, P., Taylor, C., Song, W. and Langridge, P. (1991) Cloning and characterisation of a new rye-specific repeated sequence. *Genome*, **34**, 81–7.

Guthrie, P.A.I., Magill, C.W., Frederiksen, R.A. and Odvody, G.N. (1992) Random amplified polymorphic DNA markers: a system for identifying and differentiating isolates of *Colletotrichum graminicola*. *Phytopathology*, **82**, 832–5.

Henry, R.J. (1997) Molecular and biochemical characterization of somaclonal variation, in *Somaclonal Variation and Induced Mutation in Crop Improvement* (eds S. Mohan Jain, B.S. Ahloowlia and D.S. Brar), Kluwer Academic Publishers, The Netherlands (in press).

Henson, J.M. and French, R. (1993) The polymerase chain reaction and plant diagnosis. *Annual Review of Phytopathology*, **31**, 81–109.

Holland, P.M., Abramson, R.D., Watson, R. and Gelfand, D.H. (1991) Detection of specific polymerase chain reaction products by utilising the 5'-3' exonuclease activity of *Thermus aquaticus* DNA polymerase. *Proceedings of the National Academy of Sciences, USA*, **88**, 7276–80.

Honeycutt, R. and McClelland, M. (1996) Application of the polymerase chain reaction to the detection of plant pathogens, in *The Impact of Plant Molecular Genetics* (ed. B.W.S. Sorbral), Birkhause, Boston, pp. 187–201.

Honeycutt, R.J., Sorbral, B.W.S. and McClelland, M. (1995) tRNA intergenic spacers reveal polymorphisms diagnostic for *Xanthomonas albilineans*. *Microbiology*, **141**, 3229–39.

Iqbal, M.J. and Rayburn, A.L. (1994) Stability of RAPD markers for determining cultivar specific DNA profiles in rye (*Secale cereale* L.). *Euphytica*, **75**, 215–20.

Ko, H.L. and Henry, R.J. (1994a) Rapid cereal genotype analysis, in *Improvement of Cereal Quality by Genetic Engineering* (eds R.J. Henry and J.A. Ronalds), Plenum, New York, pp. 153–7.

Ko, H.L. and Henry, R.J. (1994b) Identification of barley varieties using the polymerase chain reaction. *Journal of the Institute of Brewing*, **100**, 405–7.

Ko, H.L. and Henry, R.J. (1996) Specific 5S ribosomal RNA primers for the identification of plant species. *Plant Molecular Biology Reporter*, **14**, 33–43.

Ko, H.L., Henry, R.J., Graham, G.C., Fox, G.P., Chadbone, D.A. and Haak, I.C. (1994) Identification of cereals using the polymerase chain reaction. *Journal of Cereal Science*, **19**, 101–6.

Ko, H.L., Weining, S. and Henry, R.J. (1996) Application of primers derived from barley alpha-amylase to the identification of cereals. *Plant Varieties and Seeds*, **9**, 53–62.

Kohn, L.M., Petsche, D.M., Bailey, S.R., Novak, L.A. and Anderson, J.B. (1988) Restriction fragment length polymorphisms in nuclear and mitochondrial DNA of *Sclerotinia* species. *Phytopathology*, **78**, 1047–51.

Kolchinsky, A., Kolesnikova, M. and Ananiev, E. (1991) 'Portraying' of plant genomes using polymerase chain reaction amplification of ribosomal 5S genes. *Genome*, **34**, 1028–31.

Kolster, P., Krechting, C.F. and van Gelder, W.M.J. (1993) Expression of individual HMW glutenin subunit genes of wheat (*Triticum aestivum* L.) in relation to differences in the number and type of homologous subunits and differences in genetic background. *Theoretical and Applied Genetics*, **87**, 209–16.

Lagercrantz, U., Ellegren, H. and Anderson, L. (1993) The abundance of various polymorphic microsatellite motifs differs between plants and vertebrates. *Nucleic Acids Research*, **21**, 1111–15.

Lawson, W.R., Henry, R.J., Kochman, J.K. and Kong, G.A. (1994) Genetic diversity in sunflower (*Helianthus annus* L.) as revealed by random amplified polymorphic DNA analysis. *Australian Journal of Agricultural Research*, **45**, 1319–27.

Livak, K.J., Flood, S.J.A., Marmaro, J., Giusti, W. and Deetz, K. (1995) Oligonucleotides with fluorescent dyes at opposite ends provide a quenched probe system for detecting PCR product and nucleic acid hybridization. *PCR Methods and Applications*, **4**, 357–62.

Matthew, J., Hawke, B.G. and Pankhurst, C.E. (1995) A DNA probe for identification of *Pythium irregulare* in soil. *Mycological Research*, **99**, 579–84.

McDermott, J.M., Brandle, U., Dutly, F., Haemmerli, U.A., Keller, S., Muller, K.E. and Wolfe, M.S. (1994) Genetic variation in powdery mildew of barley: development of RAPD, SCAR, and VNTR markers. *Phytopathology*, **84**, 1316–21.

Morell, M.K., Peakall, R., Appels, R., Preston, L.R. and Lloyd, H.L. (1995) DNA profiling techniques for plant variety identification. *Australian Journal of Experimental Agriculture*, **35**, 807–19.

Morgante, M. and Olivieri, A.M. (1993) PCR-amplified microsatellites in plant genetics. *The Plant Journal*, **3**, 175–82.

Moukhamedov, R., Hu, X., Nazar, R.N. and Robb, J. (1994) Use of poly-merase chain reaction-amplified ribosomal intergenic sequences for the diagnosis of *Verticillium tricorpus*. *Phytopathology*, **84**, 256–9.

Muller, E., Brown, P.T.H., Hartke, S. and Lorz, H. (1990) DNA variation in tissue-culture-derived rice plants. *Theoretical and Applied Genetics*, **80**, 673–9.

Mullis, K.B. (1990) The unusual origin of the polymerase chain reaction. *Scientific American*, 36–43.

Muralidharan, K. and Wakeland, E.K. (1993) Concentration of primer and template qualitatively affects products in random-amplified polymorphic DNA PCR. *BioTechniques*, **14**, 362–4.

Rani, V., Parida, A. and Raina, S.N. (1995) Random amplified polymorphic DNA (RAPD) markers for genetic analysis in micropropagated plants of *Populus deltoides* Marsh. *Plant Cell Reports*, **14**, 459–62.

Riede, C.R., Fairbanks, D.J., Anderson, W.R., Kehrer, R.L. and Robinson, L.R. (1994) Enhancement of RAPD analysis by restriction–endonuclease digestion of template DNA in wheat. *Plant Breeding*, **113**, 254–7.

Roder, M.S., Plaschke, J., Konig, S.U., Borner, A., Sorells, M.E., Tanksley, S.D. and Ganal, M. (1995) Abundance, variability and chromosomal loca-tion of microsatellites in wheat. *Molecular and General Genetics*, **246**, 327–33.

Rogstad, S.H. (1994) Inheritance in turnip of variable-number tandem-repeat genetic markers revealed with synthetic repetitive DNA probes. *Theoretical and Applied Genetics*, **89**, 824–30.

Rongwen, J., Akkaya, M.S., Bhagwat, A.A., Lavi, U. and Cregan, P.B. (1995) The use of microsatellite DNA markers for soybean genotype identification. *Theoretical and Applied Genetics*, **90**, 43–8.

Rowhani, A., Maningas, M.A., Lile, L.S., Daubert, S.D. and Golino, D.A. (1995) Development of a detection system for viruses of woody plants based on PCR analysis of immobilized virons. *Phytopathology*, **85**, 347–52.

Saghai Maroof, M.A., Biyashev, R.M., Yang, G.P., Zhang, Q. and Allard, R.W. (1994) Extraordinarily polymorphic microsatellite DNA in barley: species diversity, chromosomal locations, and population dynamics. *Proceedings of the National Academy of Sciences*, USA, **91**, 5466–70.

Saiki, R.K., Gelfand, D.H., Stoffel, S., Scharf, S.J., Higuchi, R., Horn, G.T., Mullis, K.B. and Erlich, H.A. (1988) Primer-directed enzymatic amplifica-tion of DNA with a thermostable DNA polymerase. *Science*, **239**, 487–91.

Schweizer, G., Ganal, M., Ninnemann, H. and Hemleben, V. (1988) Species-specific DNA sequences for identification of somatic hybrids between *Lycopersicon esculentum* and *Solanum acaule*. *Theoretical and Applied Genetics*, **75**, 679–84.

Shewry, P.R. (1995) Plant storage proteins. *Biological Review*, **70**, 375–426.

Smith, S. and Chin, E. (1992) The utility of random primer-mediated profiles, RFLPs, and other technologies to provide useful data for varietal protec-tion in proceedings of the symposium, *Applications of RAPD Technology to Plant Breeding*, Crop Science Society of America, pp. 46–9.

Stockton, T., Sonnante, G. and Gepts, P. (1992) Detection of minisatellite sequences in *Phaseolus vulgaris*. *Plant Molecular Biology Reporter*, **10**, 47–59.

Tancred, S.J., Zeppa, A.G. and Graham, G.C. (1994) The use of PCR–RAPD technique in improving the plant variety rights description of a new Queensland apple (*Malus domestica*) cultivar. *Australian Journal of Experimental Agriculture*, **34**, 665–7.

Taylor, B.H., Young, R.J. and Scheuring, C.F. (1993) Induction of a proteinase inhibitor II-class gene by auxin in tomato roots. *Plant Molecular Biology*, **23**, 1005–14.

Taylor, G.R., Haward, S., Noble, J.S. and Murday, V. (1992) Isolation and sequencing of CA/GT repeat microsatellites from chromosomal libraries without subcloning. *Analytical Biochemistry*, **200**, 125–9.

Taylor, P.W.J., Geijskes, J., Ko, H.L., Fraser, T.A., Henry, R.J. and Birch, R.G. (1995) Sensitivity of random amplified polymorphic DNA analysis to detect genetic change in sugarcane during tissue culture. *Theoretical and Applied Genetics*, **90**, 1169–73.

Thomas, M.R. and Scott, N.S. (1993) Microsatellite repeats in grapevine reveal DNA polymorphisms when analysed as sequence tagged sites (STSs). *Theoretical and Applied Genetics*, **86**, 985–990.

Thomas, M.R., Cain, P. and Scott, N.S. (1994) DNA typing of grapevine: a universal methodology and database for describing cultivars and evaluating genetic relatedness. *Plant Molecular Biology*, **25**, 939–49.

Thomson, D. and Dietzgen, R. G. (1995) Detection of DNA and RNA plant viruses by PCR and RT–PCR using a rapid virus release protocol without tissue homogenization. *Journal of Virological Methods*, **54**, 85–95.

Torres, A.M., Millan, T. and Cubero, J.I. (1993) Identifying rose cultivars using random amplified polymorphic DNA markers. *HortScience*, **28**, 333–4.

Tsuchiya, Y., Kano, Y. and Koshino, S. (1994) Identification of lactic acid bacteria using temperature gradient gel electrophoresis for DNA fragments amplified by polymerase chain reaction. *Journal of the American Society of Brewing Chemists*, **52**, 95–9.

Tsuchiya, Y., Araki, S., Sahara, H., Takashio, M. and Koshino, S. (1995) Identification of malting barley varieties by gnome analysis. *Journal of Fermentation and Bioengineering*, **79**, 429–32.

Tyagi, S. and Kramer, F.R. (1996) Molecular beacons: probes that fluoresce upon hybridization. *Nature Biotechnology*, **14**, 303–7.

Verpy, E., Biasotto, M., Meo, T. and Tosi, M. (1994) Efficient detection of point mutations on color-coded strands of target DNA. *Proceedings of the National Academy of Sciences, USA*, **91**, 1873–7.

Vos, P., Hoger, R., Bleeker, M., Reijans, M., van de Lee, T., Hornes, M., Frijters, A., Pot, J., Peleman, J., Kuiper, M. and Zabeau, M. (1995) AFLP: a new technique for DNA fingerprinting. *Nucleic Acids Research*, **23**, 4407–14.

Ward, E. (1995) Improved polymerase chain reaction (PCR) detection of *Gaeumannomyces graminis* including a safeguard against false negatives. *European Journal of Plant Pathology*, **101**, 561–6.

Weining, S. and Henry, R.J. (1995) Molecular analysis of DNA polymorphism of wild barley (*Hordeum spontaneum*) using the polymerase chain reaction. *Genetic Resources and Crop Evolution*, **41**, 273–81.

Weining, S. and Langridge, P. (1991) Identification and mapping of polymorphisms in cereals based upon the polymerase chain reaction. *Theoretical and Applied Genetics*, **82**, 209–16.

Weising, K., Atkinson, R.G. and Gardner, R.C. (1995) Genomic fingerprinting by microsatellite-primed PCR: a critical evaluation. *PCR Methods and Applications*, **4**, 249–55.

Welsh, J. and McClelland, M. (1990) Fingerprinting genomes using PCR with arbitrary primers. *Nucleic Acids Research*, **18**, 7213–18.

Williams, J.G.K., Kubelik, K.J., Livak, K.J., Rafalski, J.A. and Tingey, S.V. (1990) DNA polymorphisms amplified by arbitrary primers are useful genetic markers. *Nucleic Acids Research*, **18**, 6531–5.

Winberg, B.C., Zhou, Z., Dallas, J.F., McIntyre, C.L. and Gustafson, J.P. (1993) Characterization of minisatellite sequences from *Oryza sativa*. *Genome*, **36**, 978–83.

Wu, K.-S. and Tanksley, S.D. (1993) Abundance, polymorphism and genetic mapping of microsatellites in rice. *Molecular and General Genetics*, **241**, 225–35.

Zabeau, M. and Vos, P. (1993) Selective restriction fragment amplification: a general method for DNA fingerprinting. EPO Patent No. 0534858A1.

Zhao, X. and Kochert, G. (1993) Phylogenetic distribution and genetic mapping of a $(GGC)_n$ microsatellite from rice (*Oryza sativa* L.). *Plant Molecular Biology*, **21**, 607–14.

Zhao, X., Wu, T., Xie, Y. and Wu, R. (1989) Genome-specific repetitive sequences in the genus *Oryza*. *Theoretical and Applied Genetics*, **78**, 201–9.

Zietkiewicz, E., Rafalski, A. and Labuda, D. (1994) Genomic fingerprinting by simple sequence repeat (SSR)-anchored polymerase chain reaction amplification. *Genomics*, **20**, 176–83.

Estimation of genetic variation in plants using molecular techniques

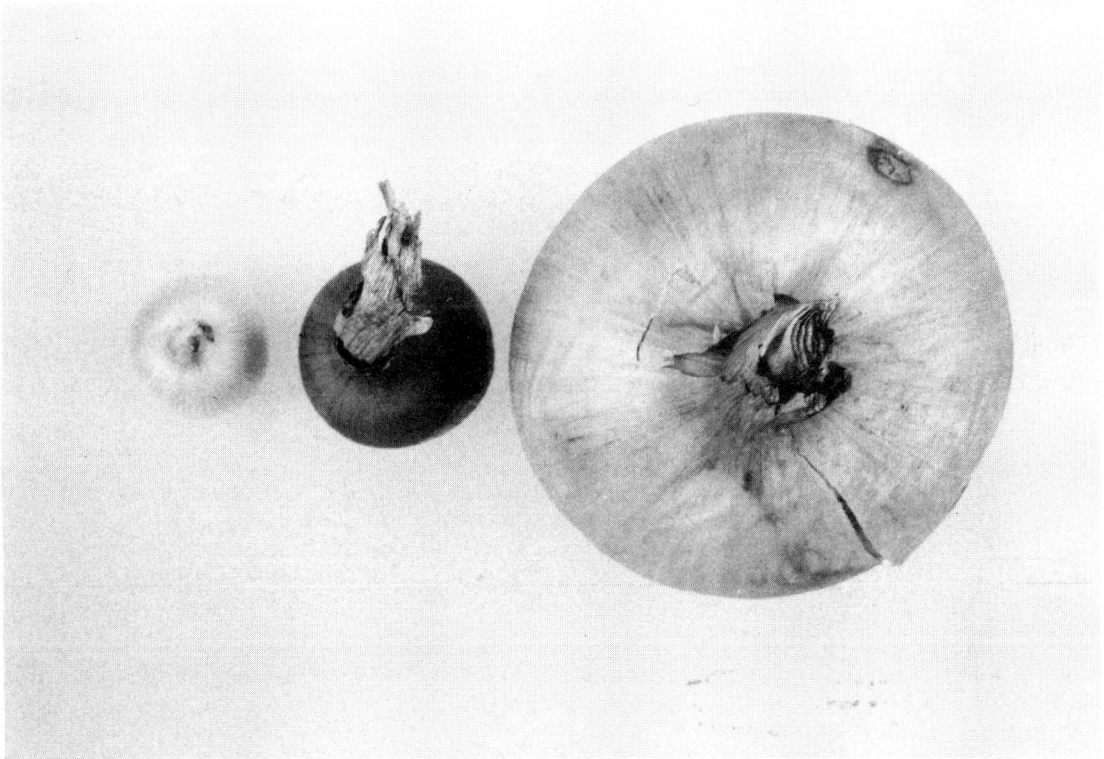

2

Genetic variation in onions (*Allium cepa* L.) These varieties vary in dry matter content and suitability for drying (Darbyshire and Henry, 1979). Molecular analysis can be used to define genetic variation in plants and to establish genetic relationships both within and between species.

After reading this chapter you should understand:

Basis of Genetic Variation in Plants ● Molecular Methods for
Estimation of Genetic Variation ● Phylogenetic Analysis
● Mitochondrial Genome ● Chloroplast Genome ● Nuclear
Genome ● Population Genetics ● Management of Wild Plants
● Conservation of Endangered Species ● Control of Weeds
● Management of Plant Genetic Resources ● Plant Genetic
Resource Collections ● Applications in Plant Improvement
● Genetic Diversity in Plant Pathogens

2.1 Introduction

The genetic variation in plants and plant populations is of considerable practical interest. Agriculture and food production depend upon
the use of highly productive plant genotypes. The conventional
breeding of crop plants is based upon selecting desirable genotypes
from the genetic variation available and manipulating all or as many
as possible of the desirable traits into one individual to develop a
commercial variety. Application of molecular biology in plant genetic
improvement (breeding) will be described in Chapter 3. This chapter
covers other applications of the analysis of genetic variation in plants.
Phylogenetic analysis, population genetics, the study and conservation
of endangered species, the management of weed populations, the establishment and management of genetic resource collections and the
analysis of genetic variation in plant pathogens will be covered in this
chapter.

The diversity of species in an environment has been shown to
contribute to the sustainability and productivity of the ecosystem
(Tilman *et al.*, 1996). Reduction of biological diversity may result in
increased losses of nutrients from the soil and lower productivity of
a plant community. Molecular techniques provide useful tools for the
study of the influence of genetic diversity of plants on the sustainability of ecosystems. This diversity may be analysed at many levels.
Biodiversity in an ecosystem is usually considered in terms of the
numbers of species. However, diversity within species may also
contribute significantly to the productivity of the system. Molecular
methods offer objective options for assessment of biodiversity and may
be the key to the development of appropriate conservation strategies
(Szmidt, 1994).

Genetic variation in plant populations may be caused and maintained by a variety of mechanisms including mutation, sexual
recombination, migration and gene flow, genetic drift and genetic
selection.

- **Mutation**: single base mutations can arise by simple base substitution or by insertion or deletion of a nucleotide. A single base substitution may alter a single amino acid in a protein or may be silent, depending on the position within the codon. The insertion or deletion of a single base can cause a frame shift and result in a major change in the protein encoded and the resultant phenotype. Changes in larger numbers of nucleotides may result from the duplication of segments, their inversion or deletion. All of these changes may occur at random and have dramatic effects or result in no change, depending on the presence of important genes in the DNA segments undergoing mutation.
- **Recombination**: mating results in the recombination of alleles in individuals and allows the spread of new alleles arising by mutations throughout the population. Plants that reproduce sexually may be inbreeding (selfing) or outbreeding (outcrossing) or may be intermediate. Outbreeding may be forced by self-incompatibility (the inability of pollen to successfully self-pollinate). Other plants reproduce asexually, eliminating recombination as a mechanism for the generation of genetic variation.
- **Migration**: genes may move from one population to another by migration. This may often occur by the dispersal of seed.
- **Genetic drift**: small populations allow random or chance changes, resulting in the loss of rare alleles.
- **Selection**: the frequency of alleles in a population may change as a result of natural selection acting on alleles that confer either an advantage or disadvantage to the survival and reproduction of the individuals in which they occur.

Analysis of the genetic variation in plants using molecular techniques can be used to investigate the extent to which plant populations have been influenced by these processes.

The evolution of flowering plants

The evolution of flowering plants from an ancestral complex is depicted in Figure 2.1. The analysis of changes in highly conserved genes can be used to calibrate a molecular clock and associate the steps in the evolution of angiosperms with the geological time scale (Figure 2.2). Angiosperms probably arose in the early Cretaceous and were widespread in the mid Cretaceous period. Many modern angiosperm families trace their origin to the late Cretaceous.

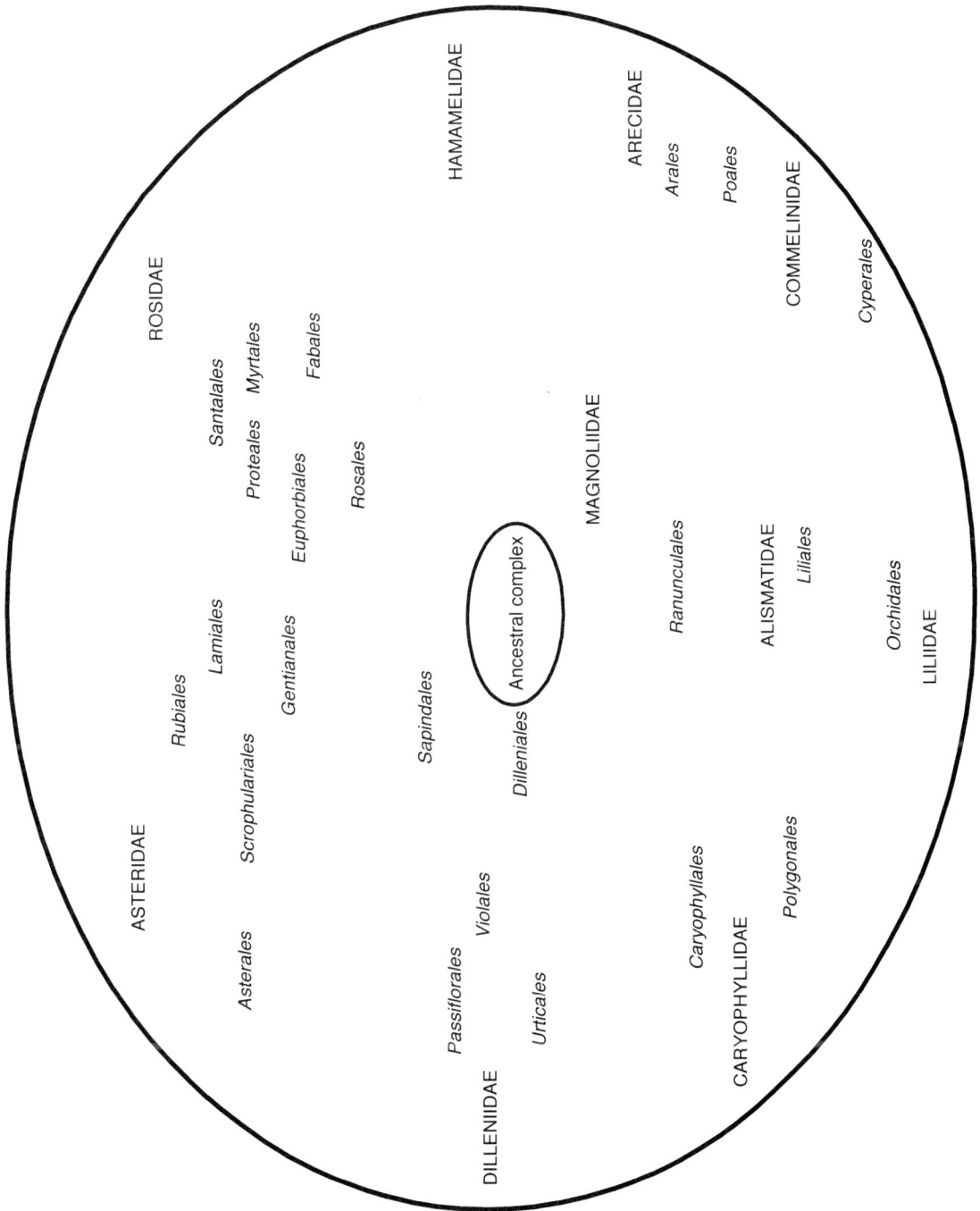

HAMAMELIDAE

ARECIDAE

Arales

Poales

ROSIDAE

COMMELINIDAE

Santalales

Myrtales

Proteales

Fabales

Cyperales

Rosales

Euphorbiales

MAGNOLIIDAE

Gentianales

Lamiales

Rubiales

Scrophulariales

Sapindales

Ranunculales

ALISMATIDAE

Liliales

Ancestral complex

Dilleniales

Orchidales

ASTERIDAE

LILIIDAE

Violales

Asterales

Caryophyllales

Polygonales

Passiflorales

Urticales

CARYOPHYLLIDAE

DILLENIIDAE

Era	Period	Epoch	Time in millions of years	Probable time of origin of plant groups
Cenozoic	Quaternary	Recent		
		Pleistocene	1.5	
	Tertiary	Pliocene	12	
		Miocene	25	
		Oligocene	34	
		Eocene	60	
		Paleocene	63	
Mesozoic	Cretaceous	Upper Lower	132	
	Jurassic	Upper Middle Lower	180	
	Triassic	Upper Middle Lower	225	
Paleozoic	Permian		275	
	Pennsylvanian	Upper Middle Lower	310	
	Mississippian	Upper Lower		
	Devonian	Upper Middle Lower	350	
	Silurian	Upper Middle Lower	405	
	Ordovician	Upper Middle Lower	430	
	Cambrian	Upper Middle Lower	485	
			600	
Pre-Cambrian			1700	
			4500	

Angiosperms · Gymnosperms · Mosses · Liverworts · Algae and fungi

Figure 2.2 Geological time scale. (From Bold, 1961; Barlow, 1981.)

2.2.1 MOLECULAR METHODS AVAILABLE

The analysis of genetic variation or diversity in plants has been traditionally assessed by analysis of morphological or biochemical traits. The assessment of phenotype may not be a reliable measure of genetic difference because of the influence of environment on gene expression. The analysis of plant DNA allows the direct assessment of variation in genotype.

The complete array of techniques used for analysis of DNA can be applied to the assessment of genetic variation in plants. These techniques may be divided into those that involve molecular marker methods and those based upon the comparison of gene sequences at specific loci. The molecular marker methods may also be used for plant identification and have been introduced in Chapter 1. The analysis of differences in sequences at specific loci may be used to develop a specific diagnostic test for identification purposes (Chapter 1) but is more commonly applied to the analysis of relative genetic distances in phylogenetic analysis. In many poorly described systems an analysis of genetic variation must precede the development of diagnostic methods to distinguish the genetic groups present. The type of molecular method used to measure genetic distances in plants will vary depending upon the magnitude of the genetic differences being assessed (Table 2.1). Techniques such as RAPD analysis may be useful for distinguishing different genotypes within a plant cultivar while sequence analysis of the ribosomal genes may allow species or higher level analysis.

The molecular techniques of DNA sequencing and molecular marker analysis are now relatively routine techniques. Some useful practical protocols are described in Chapter 5. The analysis of these data and its interpretation in the measurement of genetic relationships between different plants is more complex. The reliability of the conclusions of these studies is often dependent upon the rigour of the data analysis and interpretation. Phylogenetic studies and analysis of population genetics both require careful data interpretation. All methods require certain basic assumptions and have strengths and weaknesses (West and Faith, 1990). Mathematical methods have been developed to allow correction for the errors in estimation of genetic distance associated with the scoring of complex gels such as those generated in RAPD analysis and with the difficulty of reproducing the DNA extraction and amplification (Lamboy, 1994).

The relative information content of molecular markers can be compared by calculation of a Polymorphism Information Content (PIC; Anderson *et al.*, 1993).

2.2 Molecular methods for measuring genetic variation

Table 2.1 Molecular methods suitable for different levels of genetic distinction[†]

| | population genetics → phylogenetic analysis | | | | | |
	Individuals	Variety	Species	Genus	Family	Higher levels
SSR	**	***	*			
RAPD	**	***	**	*		
RFLP						
Nuclear genes	**	***	**	*		
Mitochondrial genes		*	**	**		
Chloroplast genes		**	***	**		
Sequencing						
ITS		**	***	**	*	
Ribosomal genes (18S)			*	**	***	***
Chloroplast genes				*	**	***
Less conserved proteins	*	**	***	**	*	
Conserved proteins			*	**	***	***

[†]Numbers of asterisks indicate increasing use of the method.

$$\text{PIC} = 1 - \sum_{j}^{n} p_{ij}^{2}$$

where p_{ij} = the frequency of the jth pattern for marker i summed over n patterns.

This index may be used to compare the value of two different marker systems in the analysis of genetic polymorphism. Roder *et al.* (1995) reported a much greater PIC for microsatellites (PIC = 0.63) in wheat compared with RFLPs (PIC = 0.30). The PIC is influenced very little by rare alleles, but provides a measure that is influenced by both the number and frequency of alleles. The maximum PIC value for a RAPD marker would be 0.5 (band present in 50% of individuals) since only two alleles are assumed in analysis of RAPDs (presence and absence of the band). The choice of a particular type of molecular marker may be determined by many factors in addition to the PIC value of the marker. For example, the cost and ease of obtaining the data and the availability of sufficient markers may justify the use of markers with a low PIC value in some applications.

2.2.2 DATA ANALYSIS

Data may be of two types, discrete characters or similarities. Discrete characters include data on sequence differences while similarity data is an estimation of the distance between two individual taxa.

Sequence alignment

The correct alignment of sequences is necessary to allow comparison of sequence differences.

The examples below show that alignment with a sequence from a hypothetical species 1. The sequence from species 2 has had gaps inserted to allow for an apparent deletion of the three bases (positions 16–18). The sequence of species 3 has a A substituted for G at position 5. Species 4 shows more differences; a deletion of bases 7–14 and substitutions at positions 23 and 24.

```
Position              10         20         30
                      |          |          |
species 1   GATCGACCTACGATCGAACGGCTTATCTGT

species 2   GATCGACCTACGATC———————CGGCTTATCTGT

species 3   GATCAACCTACGATCGAACGGCTTATCTGT

species 4   GATCGA———————————————-CGAACGGCACATCTGT
```

Analysis of sequence data requires alignment of the sequences and this may require the insertion of gaps. This allows the identification of positional homologies. Algorithms are available to allow computerized alignment of sequence. One approach is to score matches as positive, mismatches as zero, and gaps as negative, and to align to achieve a maximum score. Gaps must be weighted more heavily than mismatches to ensure that all mismatches are not eliminated by excessive insertion of gaps. Specific methods have been developed for calculation of evolutionary distances based upon the rates of base substitution (Kimura, 1980).

Similarity data define the differences between pairs of taxa. Character data can be transformed to similarity data for analysis.

As an example, estimation of similarities based upon RAPD results for two samples A and B can be conducted as follows:

$$\text{Similarity (F)} = 2(n_{xy})/n_x + n_y$$

where n_{xy} = number of bands in common to sample A and sample B; n_x = number of bands for sample A; and n_y = number of bands for sample B.

If sample A and B both resulted in the amplification of a total of 100 bands (pooling data from analysis with 12 primers) with 75 of them being common, then $F = 2 \times 75/100 + 100 = 0.75$ (or 75% similarity).

The above equation was proposed by Nei and Li (1979), an alternative being Jaccard's coefficient:

$$F' = n_{xy} / (n_t - n_z)$$

where n_{xy} = number of bands in common to sample A and sample B; n_t = total number of bands present in all samples; and n_z = number of bands not present in sample A or B but found in other samples.

In the example above, if other samples contributed to give a total of 200 bands amplified with 75 of them being absent in both A and B, then the Jaccard coefficient $F' = 75/200–75 = 0.66$.

F may also be calculated in other ways (Virk *et al.*, 1995b):

$$F'' = (n_{xy} + n_x)/n_t$$
$$F''' = n_z/ (n_t - n_{xy})$$

Dissimilarity is usually defined as 1 minus similarity (1–F). A distance matrix of similarity or dissimilarity can be produced for any set of individuals.

If a third sample, C (also resulting in 100 bands: 50 in common with A and 60 in common with B) is considered, the data could be represented in the following dissimilarity matrix:

Generation of a matrix
of 1–F (dissimilarity) values:

	A	B	C
A	1.00	0.25	0.50
B	0.25	1.00	0.40
C	0.50	0.40	1.00

A phenogram can be generated using an unweighted pair group method with arithmetic mean analysis (UPGMA) as follows.

If we consider the closest pair of samples (A and B) as a cluster, the average distance of A and B from a common ancestor can be considered to be: $0.25/2 = 0.125$. We then treat A and B as a single entity AB. The dissimilarity of AB from C (the next closest sample) is now calculated as the average of the dissimilarities from A and B.

$$0.5 + 0.4/2 = 0.45.$$

In the cluster of C and AB, the average distance of C and AB from a common ancestor can be considered to be: $0.45/2 = 0.225$.

Phenogram for samples A, B and C:

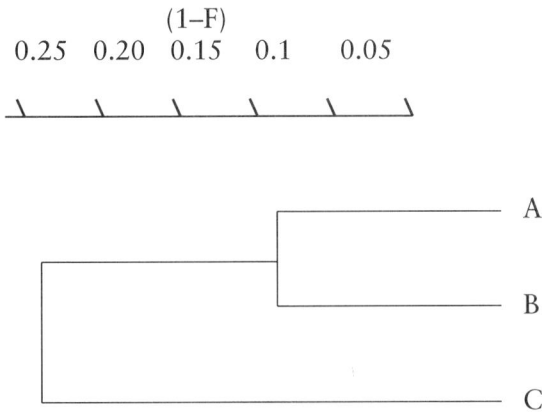

```
               (1–F)
   0.25   0.20   0.15    0.1       0.05

   \        \       \        \        \           \
```

```
                          ┌──────────────── A
                    ┌──────┤
             ┌──────┤      └──────────── B
             │      │
             └──────────────────── C
```

This case is much simpler than most because only three samples were involved. The construction of trees based upon molecular distance data is usually conducted using computer software; by the unweighted pair-group method with arithmetic averaging (UPGMA) using NT–SYS, the neighbour-joining method or maximum likelihood calculated using PHYLIP (Phylogenetic Inferences Package) or for character data using Wagner parsimony with PAUP (Phylogenetic Analysis Using Parsimony) or MacClade software.

Cladistic analysis attempts to recreate true phylogeny by identifying individuals with shared characters. These individuals can be considered as a clade arising from a common ancestor. Many phylogenetic trees can be created by this process. The tree with the smallest number of steps (the most parsimonious) is considered the most likely to represent the true phylogeny. A consensus tree retaining the most common branches may be selected from the most parsimonious trees. Statistical tests such as bootstrapping or jack-knifing may be used to evaluate the results. Bootstrapping involves random resampling from the original data set and analysis of the data 100 to 1000 times to establish that the result is robust. Jack-knifing resamples the data by dropping out different samples and reanalysing.

An example of a phylogenetic tree generated for 18 gymnosperm samples is given in Figure 2.3 (Graham et al., 1996). The tree was generated by analysis of the entire 1700 bp of the 18S rRNA sequences by the standard Wagner parsimony method using PAUP. The numbers on the branches indicate the percentage of bootstraps where the branch was observed.

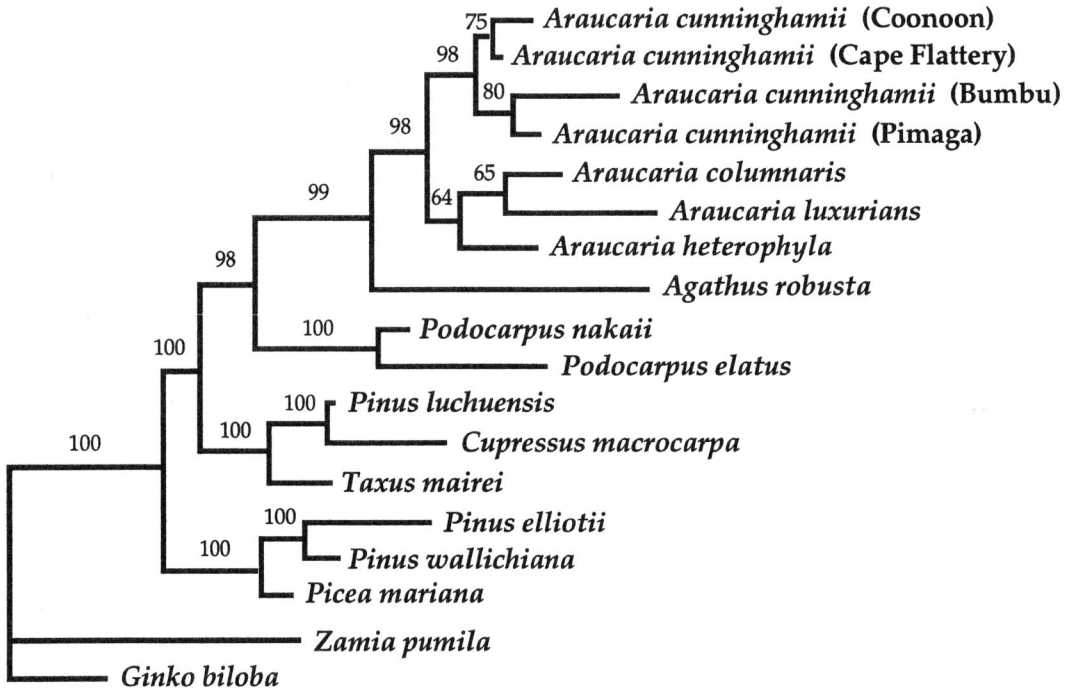

Figure 2.3 Most parsimonious tree generated by analysis of 18S rRNA sequences from gymnosperms. Numbers indicate the percentage of bootstraps in which the branch was observed.

<table>
<tr><td>

2.3 Phylogenetic analysis of plants using molecular techniques

</td></tr>
</table>

2.3.1 INTRODUCTION

Phylogenetic analysis attempts to define the evolutionary history of plants. Current plant populations are grouped in an attempt to explain their development from ancestral taxa. The practical value of such analysis is primarily associated with the use of such information in plant genetic improvement. Classification of plants in systematic botany (Cronquist, 1981) is largely based upon assumptions of phylogeny. An example of the hierarchy of categories into which higher plants are classified is given in Table 2.2. Knowledge of genetic relationships between plant-based taxonomic information can provide clues to potential genetic resources for use in crossing or in the isolation of useful genes. A complete listing of the families of flowering plants is given in Appendix B.

Data defining genetic distances between taxa are often presented in the form of a matrix of similarities or dissimilarities. These data can be used to generate phylogenetic trees and to deduce pathways of evolution of higher plants. Cladograms show the evolutionary relationship between taxa in the form of a tree. Trees can be rooted (trees

Table 2.2 Classification of a plant showing the hierarchy of taxonomic categories. The example given is the Golden Wattle, as described by Kanis (1981)

REGNUM (kingdom)	*Eukaryota* (organisms of nucleate cells)
Subregum	*Embryobionta* (stem plant)
DIVISIO (division)	*Magnoliophyta* (flowering plants)
Subdivisio	
CLASSIS (class)	*Magnoliopsida* (Dicotyledons)
Subclassis	*Rosidae*
ORDO (order)	*Fabales* (leguminous plants)
Subordo	
FAMILIA (family)	*Mimosaceae* R. Br.
Subfamilia	
Tribus (tribe)	*Acacieae* (Reichb.) Endl.
Subtribus	
GENUS	*Acacia* Miller
Subgenus	*Acacia* subg. *Heterophyllum* Vassal
Sectio (section)	Acacia sect. *Phyllodineae*
Series	*Acacia* ser. *Uninerves* Benth.
Subseries	*Acacia* subser. *Racemosae* Benth.
SPECIES	*Acacia pycnantha* Benth.
Subspecies	
Varietas (variety)	*Acacia pycnantha* Benth. var. *pycnantha*

for which a common ancestor is identified) or unrooted (trees for which the common ancestor is not identified). The analysis of such data (see pp. 64–67) is difficult and sometimes controversial.

2.3.2 ANALYSIS OF MITOCHONDRIAL DNA

Plant mitochondrial genomes are larger and more variable in size than those from animals, with sizes ranging from 100 to 2500 kb. The presence of multiple copies of the mitochondrial genome in each cell makes the mitochondria an abundant source of DNA for analysis of phylogenetic relationships. Some mitochondrial genes originated in chloroplasts (Hirai and Nakazono, 1993) as shown in Figure 2.4. Mitochondrial genome analysis has been widely used in plant systematics (Crozier, 1990). RFLP analysis of chloroplast, mitochondrial and nuclear genomes in two rice species suggested that the rates of nuclear and mitochondrial evolution were similar while the chloroplast variability was less than half of that in the other two genomes (Ishii *et al.*, 1993).

Early studies used RFLP to investigate variation in mitochondrial genes. More recently, direct sequencing of genes amplified by PCR has

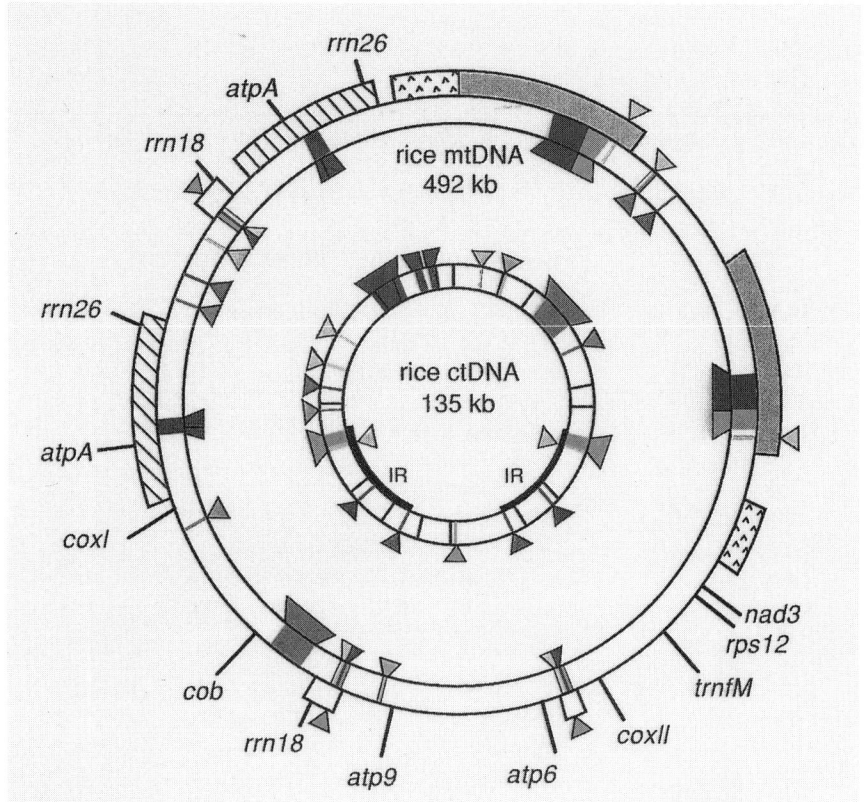

Figure 2.4 Mitochondrial genome of rice showing elements derived from the chloroplast. (From Hirai and Nakazono, 1993.)

been applied to comparison of mitochondrial genes. Variations in the rate of evolution of different mitochondrial genes results in the potential to select genes for any particular level of phylogenetic analysis. For example, ribosomal genes from mitochondria may be used in phylogenetic analysis (Hillis and Dixon, 1991).

Geographic origin of the wild relatives of the cultivated rubber trees (*Hevea brasiliensis*) was indicated by RFLP analysis of mitochondrial DNA (Luo *et al.*, 1995). Similar analysis of chloroplast DNA from this group revealed comparatively little polymorphism. In other cases mitochondria are not sufficiently discriminating. RFLP of mitochondrial DNA from finger millet (*Eleusine coracana*) allowed the distinction of only three groupings in 26 lines (Muza *et al.*, 1995).

Sequencing of the mitochondrial LrRNA gene has been used to establish relationships in the ectomycorrhizal fungi associated with a group of epiparasitic plants (Monotropoideae) (Cummings *et al.*, 1996). This analysis demonstrated that these fungi are highly specialized with some

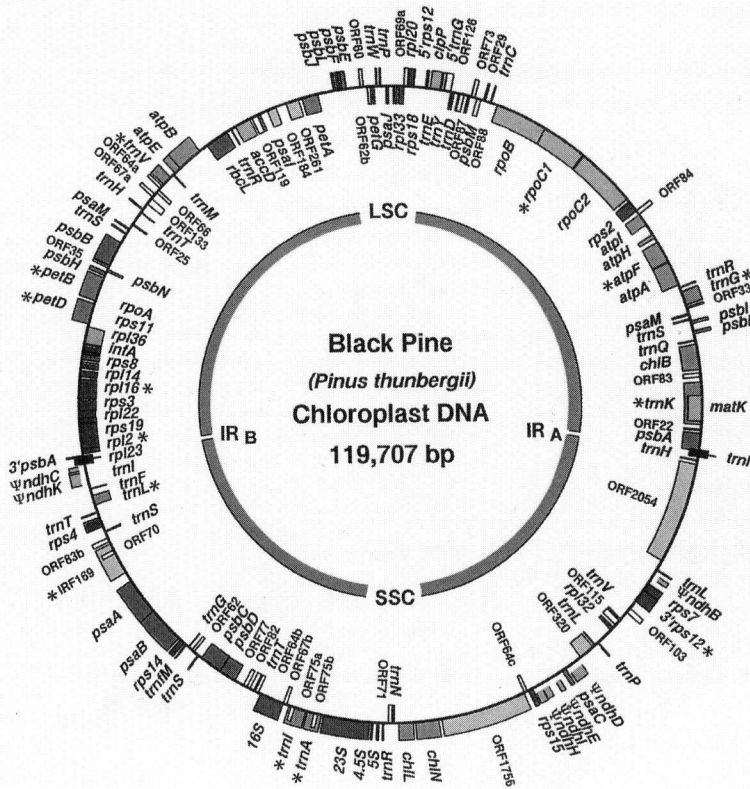

Figure 2.5 Chloroplast genome of black pine. (From Wakasugi *et al.*, 1994.)

plants being found to associate exclusively with a single species of the fungus.

2.3.3 ANALYSIS OF CHLOROPLAST GENES

The chloroplast has an additional genome unique to plants that can be used for estimation of genetic differences (Zurawski and Clegg, 1987). Many copies of this small (120–220 kb) genome (Figure 2.5) are present in each cell. Chloroplast genome analysis allows evaluation of higher level taxonomic groups because of the relatively slow rate of evolution of the chloroplast. The large subunit of ribulose 1,5-bisphosphate carboxylase (*rbc*L) and the β and ε subunits of ATPase (*atpBE*) has been used extensively as a basis for estimation of genetic relatedness. Ribosomal genes from chloroplasts are similar to those from prokaryotes and have also been useful in phylogenetic analysis.

The highly conserved nature of chloroplast genomes makes it easy to compare homology.

RFLP analysis of chloroplast DNA is useful in distinguishing species within genera (Doyle and Doyle, 1991). Chloroplast DNA from the fast-growing *Populus* tree species show intraspecific and interspecific RFLPs (Rajora and Dancik, 1995a,b) that may be useful in identification of species and hybrids (Rajora and Dancik, 1995c). Similar analysis has been used to evaluate the pathways of domestication of potatoes (Hosaka, 1995). RFLPs in chloroplast DNA have also been useful in the analysis of relationships between taxa in the Brassicaceae at the subtribe level (Warwick and Black, 1994).

Deletions in plastid genomes have been observed in long-term culture of plant cells (Kawata *et al.*, 1995). These deletions were associated with changes in plastid morphology and may be associated with loss of totipotency of plant cells in culture.

Restriction analysis of chloroplast DNA was used by Bolger and Simpson (1995) to resolve difficult problems of phylogeny in the Liliaceae and related families. The Liliaceae is a major monocot family variously considered to include genera otherwise included in the Agavaceae or Amaryllidaceae. Two major clades, one including the Agavaceae and some other taxa such as Xanthorrhoea (all with capsular fruits) and the other including the Convallariaceae, Dracenaceae and Nolinaceae (all with berries or nutlets as fruits), were distinguished in this study.

2.3.4 ANALYSIS OF NUCLEAR GENES

The nuclear genomes of plants are much larger (10^8 to 10^{11} bp) than those of the organelles and provide a vast array of possible genes for analysis (Dean and Schmidt, 1995). The genomes of plants may vary 1500-fold in size. Plants with the smallest of these genomes are successful flowering plants. The additional DNA in plants with much larger genomes may be largely non-genic but is not necessarily without function. The DNA of the nucleus is found in chromosomes. Analysis of nuclear genomes may involve comparison of specific loci, the arrangement of the loci along the chromosome and the distances between the loci. Genetic maps have been produced for the nuclear genomes of many plant species. These genetic maps do not necessarily correspond to physical maps. The genetic distance (or linkage) between two markers is determined by the frequency of recombination between the two sequences being probed but this may correspond to very different physical distances for different pairs of markers. Comparative mapping of plant species has shown a large degree of colinearity in gene order for many related species. The analysis of plant nuclear genes provides a vast

Nuclear genomes

Nuclear genes are arranged in chromosomes. Genetic maps of plant chromosomes usually depict the haploid genome, as in the hypothetical map shown below. Markers are separated by genetic distances. Importantly, these do not necessarily relate to physical distances along the chromosome. The position of the centromere (marker C) is not always known with confidence. In the example given, 40 markers are mapped on four chromosomes. Note that the coverage is not normally even. For example, a large gap exists between markers A and B on chromosome 1 and chromosome 4 has the most detailed map.

| Chromosome 1 | Chromosome 2 | Chromosome 3 | Chromosome 4 |

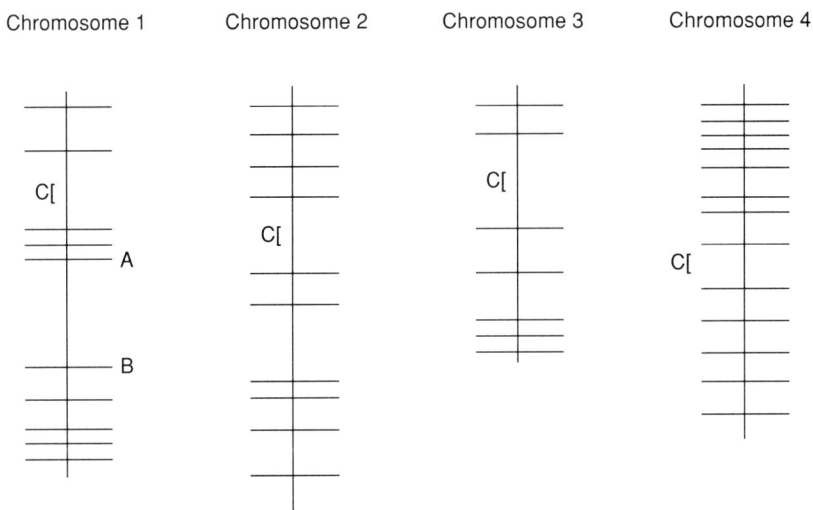

array of options for investigation of genetic differences, including the analysis of sequence differences at specific genetic loci or analysis of differences in genome arrangement. Multicopy genes such as those encoding ribosomal RNA have been very useful in phylogenetic analysis because of the presence of highly conserved sequences (permitting specific and reliable detection of these loci in different taxa) and their high copy number (providing good sensitivity for detection).

(a) Ribosomal RNA genes

Ribosomal RNA gene analysis has been the most common approach to the generation of molecular phylogenies from nuclear genes in plants (Zimmer *et al.*, 1989). The ubiquitous occurrence of ribosomal genes makes them ideal candidates as a universal molecular tool for phylo-genetic work (Solignac *et al.*, 1991). The internal transcribed spacers (ITS) of nuclear RNA genes have been found to be especially useful (Baldwin, 1992).

Phylogenetic analysis based upon ITS sequences has proven gener-ally useful in such important plant groups as the Poaceae (grasses) which includes the major food crop species (Hsiao *et al.*, 1995). Phylogenetic relationships within the *Sorghum* genus and other related genera have been established by analysis of ITS (Sun *et al.*, 1994). The ITS of *Sorghum bicolor* was 588 bp in length with a small number of insertions or deletions (four) and a much larger number of base substi-tutions (126) in the 17 species compared. Other examples of the use of ITS sequences in phylogenetic analysis include the study of members of the Asteraceae (daisy family) (Baldwin, 1992) and the Betulaceae (alders and birches) (Savard, 1993).

One difficulty with this approach is that more than one ITS sequence may be found in a species or even within an individual. Ritland *et al.* (1993) found more than one type of ITS in species from the common yellow monkey flower, *Mimulus guttatus*, complex. Two of the three ITS types present were found in more than one species, suggesting that they were present before speciation. These studies indicate the impor-tance of ensuring that the same type of ITS sequence is being compared across taxa in any phylogenetic analysis. Cummings (1992) designed primers for PCR of ribosomal genes that were specific for the amplifi-cation of plant ribosomal sequences in roots infected with mycorrhiza. These primers allow the specific identification of the plant host species. This type of approach may be important in the analysis of apparently uninfected plant material when PCR is being used to amplify ribosomal genes for sequencing (B. Congdon, personal communication) because of the risk of amplifying genes from undetected fungal contaminants.

The spacers between the ribosomal genes of wheat have been shown to exhibit variation (May and Appels, 1987). The polymorphisms may be associated with insertions or deletions of extra repeat units in the spacers. The 5S ribosomal gene spacers are also polymorphic between wheat varieties (Cox *et al.*, 1992). Analysis of these ribosomal gene variations may allow distinction of major genetic groupings within plant species. Study of ribosomal genes of *Vicia* (Leguminosae) species by RFLP analysis (Raina and Ogihara, 1995) has indicated relation-ships between 49 species and suggests that none of them is closely related to broad bean (*Vicia faba*).

(b) Phylogenetic inference using RAPD

The RAPD method allows data for a large number of alleles to be easily collected and has been applied to many systems (Demeke and

Adams, 1994). The RAPD markers are generally assumed to be of nuclear origin, but this may not always be the case. The amount of variation detected suggests that this approach is likely to be most useful for analysis at the species or lower levels or at least the subgeneric level. RAPD markers have been used to analyse genetic relationships in *Lens* (Abo-elwafa *et al.*, 1995). The variation within cultivated lentil (*Lens culinaris* ssp. *culinaris*) was lower than that in wild relatives. *Lens culinaris* ssp. *orientalis* was indicated as the most likely wild progenitor for lentils. RAPD has also been applied to the analysis of cocoa (*Theobroma cacao* L.) populations, revealing greater variation within populations than between populations (Russell *et al.*, 1993). RAPD and RFLP markers were shown to separated 106 cocoa genotypes into the same three groups (N'Goran *et al.*, 1994). These studies have assisted in the investigation of the origins of cultivation of cocoa (de la Cruz *et al.*, 1995).

2.4 Applying molecular techniques to plant population genetics

Plant population genetics can be used to assess factors such as selection pressures in the local environment, rates of movement of genes between subpopulations and genetic drift associated with evolutionary change. An understanding of these factors can be crucial to effective management of wild plant populations. Two practical examples, the conservation of endangered species and the management of weeds, will be described in the next section (pp. 79–82).

Plant population genetics requires the calculation and analysis of allele frequencies in populations. In the absence of selection, drift or migration, a randomly mating plant population will exhibit stable allele frequencies at Hardy–Weinberg Equilibrium (HWE).

Nuclear loci in a diploid population will have a binomial distribution of allele frequencies at HWE. For example, alleles A and B present at frequencies a and b are expected to result in the following genotype frequencies at HWE:

$$AA = a^2 \qquad AB = 2ab \qquad BB = b^2$$

If for example allele A was found in 90% of the population ($a = 0.9$) and allele B in 10% of the population ($b = 0.1$) then the frequencies of genotypes at HWE would be as follows:

AA = 81%;   AB = 18%; and BB = 1%

This illustrates the higher frequencies expected for heterozygotes compared with homozygotes for alleles that are rare in the population. Tests for significant deviation from the allele frequencies predicted for HWE are used to establish that the population is not at equilibrium and the extent to which factors such as selection or migration may be occurring. At HWE, allele frequencies remain constant through successive generations. Molecular analysis can be used to show that selection for a specific trait may result in geographic heterogeneity in allele frequencies at the locus under selection compared with other loci (Taylor *et al.*, 1995).

Plants may be inbreeding, mating only with closely related individuals or in the extreme only by self-fertilization, or outcrossing, mating with unrelated individuals without restriction. Inbreeding results in a reduction in the frequency of heterozygotes in the population. Each generation of self-fertilization halves the proportion of heterozygotes in the population.

In a population where the alleles A and B are present at frequencies *a* and *b*, an inbreeding coefficient (*F*) can be defined with H representing the frequency of heterozygous genotypes.

$$F = (2ab - H)/2ab$$

At HWE no inbreeding occurs and $F = 0$ since $H = 2ab$, while in the case of repeated selfing, $F = 1$ since $H = 0$.

An analysis of allele frequencies in this way can be used to assess the extent of inbreeding in plant populations. One effect of inbreeding is to increase the chance that deleterious rare recessive alleles may become homozygous, reducing the fitness of the plant population.

The techniques that have been favoured for population genetics are those that detect a relatively large number of readily distinguishable alleles from well-defined loci. Classical analysis has been based upon allozymes (Lynch and Vaillancourt, 1995) but is being replaced by DNA-based methods (Campbell *et al.*, 1995).

The large numbers of markers that can be generated by techniques such as RAPD analysis make this an attractive method for the analysis of the genetics of plant populations (Lynch and Milligan, 1994). The application of RAPD analysis to plant population genetics is complicated by several factors. The dominant nature of RAPD markers (and other markers such as AFLPs) makes it impossible to distinguish homozygotes from heterozygotes, preventing estimation of inbreeding from the measurement of heterozygosity. This reduces the accuracy of estimation of allele frequencies relative to that possible with a co-dominant marker. Other assumptions required for the use of RAPD markers in this way are that all markers are from separate loci and that there is only one amplifiable allele per locus. Errors may also result from coincidence of different alleles giving markers of the same or indistinguishable sizes. However, this error is small and an estimate of the probability of coincidence based upon the resolution of the gel could be used to correct for this possibility. These considerations mean that the use of RAPD or AFLP data in population genetics requires the sampling of many more alleles than a conventional co-dominant marker to reach conclusions with the same degree of confidence. The

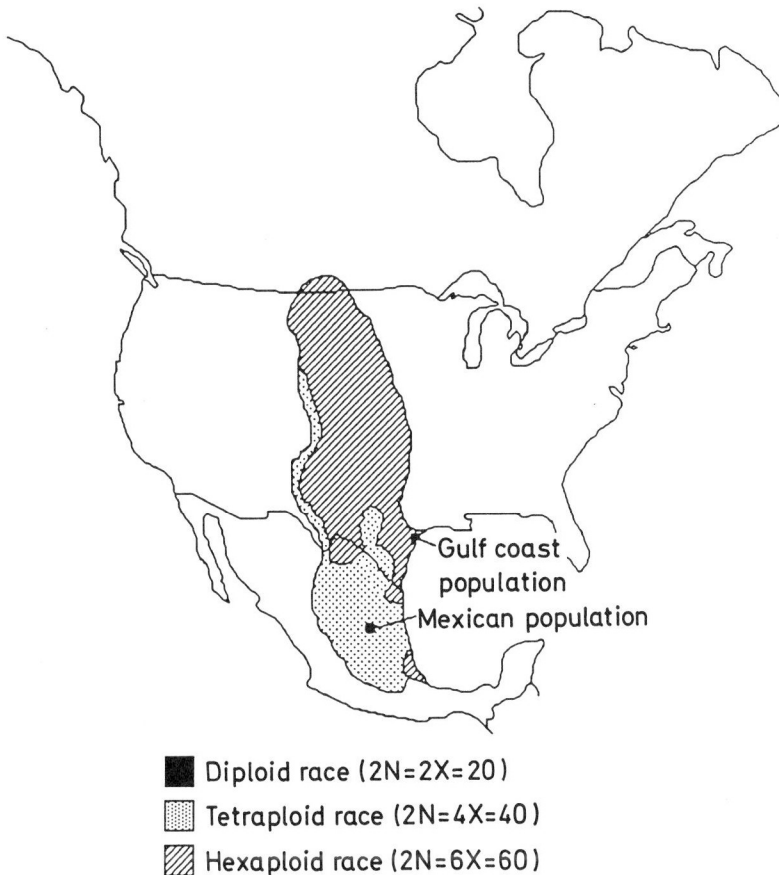

Diploid race (2N=2X=20)
Tetraploid race (2N=4X=40)
Hexaploid race (2N=6X=60)

Figure 2.6 Distribution of buffalograss (*Buchloe dactyloides*) populations in North America. (From Peakall *et al.*, 1995.)

generation of this larger amount of data may still be attractive because of the comparative simplicity of the techniques (Yu and Pauls, 1994). Peakall *et al.* (1995) compared allozymes and RAPDs for the analysis of variation within and between populations of buffalograss (*Buchloe dactyloides* (Nutt.) Engelm.) and obtained generally similar results with the two techniques. Populations from Texas and Mexico (Figure 2.6) showed considerable within- and between-population variation (Figure 2.7).

The population genetics of *Pyrenophora teres*, the causal organism of net blotch of barley, has been analysed using RAPD markers (Peever and Milgroom, 1994). Within- and between-population variability in allele frequency could be partitioned. The technique could also be used

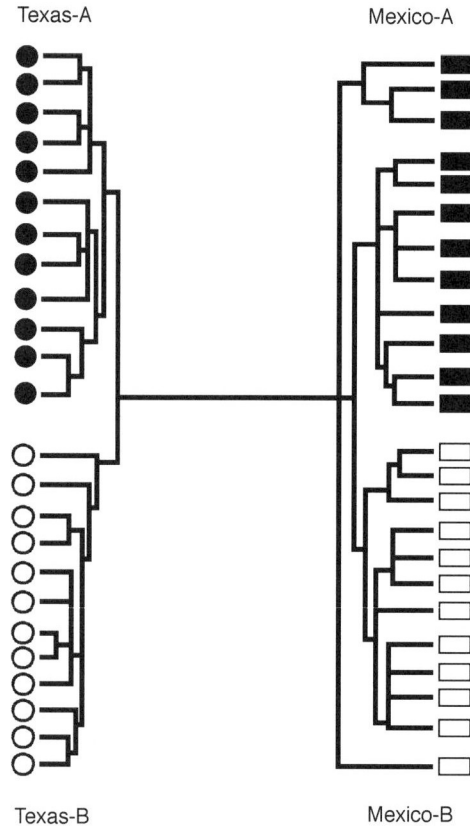

Figure 2.7 Tree relating buffalograss populations based upon analysis of RAPD data. (From Peakall *et al.*, 1995.)

to distinguish populations that may be reproducing sexually in areas of continuous barley production from a population in a region in which barley is not usually grown in the same field in consecutive seasons and that may reproduce by largely asexual processes. The population suspected of asexual reproduction showed significant gametic dysequilibrium (non-random association of RAPD loci).

The variation within and between populations of buffalograss (an outcrossing species) has been partitioned by analysing RAPD data using an Analysis of Molecular Variance (AMOVA) technique (Huff *et al.*, 1993). This study showed that within-population variation was larger and among-population variation smaller in Mexico than in Texas. The largest differences were between the regions.

Plant mitochondria are usually maternally inherited. Hence, mitochondrial gene analysis allows maternal and paternal gene flow to be distinguished.

Sampling is a key issue in the successful analysis of plant populations with molecular techniques. Analysis of too many will waste resources, while analysis of too few will not allow a satisfactory conclusion to be reached (Baverstock and Moritz, 1990).

2.5.1 INTRODUCTION

Molecular analysis of genetic variation in populations of rare or endangered plants allows management of the plant populations so as to preserve genetic diversity and may help to ensure the long-term survival of the species. A similar understanding of weed populations may be used to manage for a minimum population size. Speciation of plants involves the development of differences in plant populations that result in reproductive isolation. Molecular markers have been used to show that major genes may contribute substantially to the process of speciation. Two species of monkeyflowers (*Mimulus*) are reproductively isolated by preference for different pollinators, bumble bees and humming birds. Molecular markers (RAPDs) have been used to map quantitative trait loci accounting for at least a quarter of the variation in eight floral traits associated with pollinator preference in these species (Bradshaw *et al.*, 1995). Thus, molecular markers may be used to study major genes contributing to speciation in wild plant populations. Molecular markers may be useful in the management of plant populations composed of species that are either endangered or unwanted (e.g. weeds). For example, the maintenance of rare plant communities may require the control of the population structure of abundant or dominant species so as to prevent succession to a less desirable or common community.

> **2.5 Applying molecular techniques to the management of wild plants**

2.5.2 EXAMPLES

(a) Conservation of endangered plant species

A knowledge of the genetic variation in rare (Figure 2.8) or endangered plant species can be an important tool in management to ensure survival of the species and its genetic diversity. The relationship between geographically isolated populations or individuals may indicate the extent of genetic isolation and indicate strategies to conserve variations in wild populations by protection of appropriate habitat variation. Introductions of new individuals produced by human propagation can be managed to protect local genetic identity. Sometimes a geographically isolated population may be unique and justify conservation measures, while in other cases molecular evidence may suggest that the population is not sufficiently different to require special protection. Occasionally,

Figure 2.8 Rare plants may be better protected with the knowledge of population structure that can be obtained by molecular analysis. The plant depicted is *Banksia praemorsa* from the south coast of Western Australia.

Table 2.3 Classification of rare or endangered plants. After Leigh *et al.,* 1984.)

Category	Description
Extinct	No longer known to exist (may be defined on the basis of not having been collected for a period of time, e.g. in the past 50 years)
Endangered	In danger of extinction
Threatened or vulnerable	Likely to move into the endangered category if current causal factors continue
Rare	Small world populations (may be localized or thinly spread over a larger area)

Rare plants

Banksia praemorsa Andrews 2VC (see Figure 2.8) is restricted to coastal cliffs close to Albany in Western Australia. The species is locally abundant but found in only a small area. Conservation of a very specific habitat is essential to the survival of such species.

---

species that are considered very rare may be identified as hybrids between two common species by molecular analysis.

The classification of endangered plants into different categories (Table 2.3) may be aided by appropriate molecular analysis. Molecular analysis provides an objective basis for extending this concept to individual plant populations within the species.

Rare plants may show less genetic variation than widespread taxa. Small populations may lose rare alleles and this may be evident before loss of total gene diversity is apparent.

Fragmentation of wild plant populations may lead to a loss of genetic diversity because of genetic bottlenecks, the reduction of gene flow between the remnant populations, increased levels of inbreeding within remaining populations, and a higher incidence of genetic drift. The lower genetic diversity will leave the population less well equipped to survive changing environmental pressures and may eventually lead to extinction. Molecular techniques can be used to monitor these processes by evaluating the loss of rare alleles and the level of heterozygosity in the population.

Molecular analysis provides an important tool for the characterization of biodiversity. Identification of areas rich in endemic genotypes is an important step towards habitat conservation and management for reduced rates of species extinction (Pimm *et al.*, 1995).

Molecular analysis of very rare plant populations may be an extremely valuable tool in management of their survival. In extreme cases, if only one individual remains, phlyogenetic analysis of the plant's relationship to other taxa is the only possibility. This will indicate the uniqueness and thus the conservation value of the species. If more than one individual exists the analysis of the genetic structure of the population is possible. This will indicate if all individuals are clonal (depending on the reproductive biology) or if genetic variation exists. If all the individuals are genetically identical, the options are the same as with the existence of only one individual. If even small genetic variation exists, molecular analysis will indicate management strategies by which such variation can be preserved. Propagation from several or many individuals representing the genetic variability in the species can be attempted or strategies to protect individuals of unique

Figure 2.9 Weed populations. *Senecio madagascariensis* in Laminton National Park, Australia. Molecular methods allow distinction of introduced weed from native *Senecio* populations.

genotypes and allow their reproduction *in situ* can be implemented.

(b) Control of weeds

Weeds (Figure 2.9) are unwanted plants and frequently originate outside the environment in which they are a problem. The geographical origin of a weed species can be established by using molecular methods to compare the weed population with possible source populations. This approach may be useful in the targeting of biological control agents (e.g. insects or other pathogens) to specific weed genotypes and their selection from appropriate sources. The genetic variation in the weed population can be used to establish the frequency or number of introductions of new individuals or genotypes. Molecular analysis may thus be a good general approach to generating data useful in the management of weed populations.

Sequencing of ITS has been used to establish that fireweed (*Senecio madagascariensis* Poir.) was introduced into Australia and is distinct from closely related endemic species (L. Scott and B. Congdon, personal communication). Hybridization between the introduced species and native species has been indicated, raising possible difficulties in the control of this weed.

2.6.1 INTRODUCTION

The number of plant species has been estimated to approximate a quarter of a million. The conservation of genetic resources for this number of species is a daunting task. A relatively small number of these species have been utilized by humans (Figure 2.10) and a very much smaller number are cultivated (Table 2.4 gives world production levels for the major crops). Three species – wheat, rice and maize – represent about half of all human food. Maintenance of genetic resources for these major crops is vital to the security of human food supply. Molecular techniques provide opportunities to better conserve and manage this essential resource. Molecular techniques may also be

> **2.6 Using molecular techniques to manage plant genetic resources**

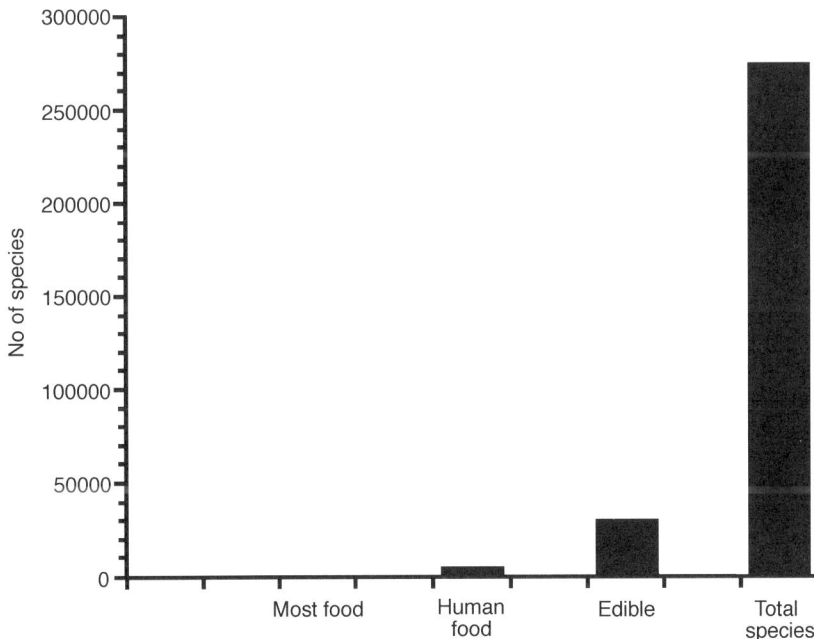

Figure 2.10 Unexploited plant genetic diversity. The diagram illustrates the very small proportion of plant species that are exploited by humans for food.

Table 2.4 World production of major crops. (From FAO, 1991.)

Crop	1990 production (1000 tonnes)	Crop	1990 production (1000 tonnes)
Wheat	601 723	Apples	40 518
Rice	521 703	Nuts	4 379
Maize	479 340	Oil crops	74 945
Barley	181 946	Sugar	110 823
Potatoes	268 107	Coffee	6 282
Pulses	58 846	Tea	2 533
Vegetables and melons	450 986	Cotton	18 447
		Jute	3 669
Grapes	59 873	Tobacco	7 076
Citrus	72 998	Rubber	4 922
Bananas	46 923		

Table 2.5 Seed stored in international agricultural centres. (From Chin, 1994.)

Centre	Crop	No. of accessions
AVRDC (Asian Vegetable Research and Development Centre)	Vegetables	35 948
CIAT (Centro Internacional de Agricultura Tropical, Columbia)	Beans Tropical forages	26 852 22 818
CIMMYT (Centro Internacional de Mejoramiento de Maizy Trigo, Mexico	Cereals	86 514
ICARDA (International Centre for Agricultural Research in the Dry Areas, Syria	Cereals Food legumes Forages	47 907 20 092 19 771
ICRISAT (International Crop Research Institute for the Semi-Arid Tropics, India)	Cereals Legumes	61 891 41 194
IITA (International Institute of Tropical Agricultural, Nigeria)	Legumes Cereals	20 200 13 700
IRRI (International Rice Research Institute, Philippines)	Rice	86 402

Figure 2.11 Plant genetic resource collections. The genetic resource collection in Tsukuba, Japan.

used to make use of the vast reservoir of currently unexploited plant genetic resources in inedible or otherwise unused species.

2.6.2 PLANT GENETIC RESOURCE COLLECTIONS

Most of our *ex situ* plant genetic resources are in the form of seed collections (Table 2.5). Seedbanks have been estimated to contain a total of 3.5 million seed accessions contained in about 1000 collections in 121 countries (Chin, 1994). The advantage of seed collections is the generally long life of seeds when carefully stored, usually at low temperatures and moisture contents. This approach is not suitable for all plant species. Recalcitrant species are killed by the low temperatures and drying required in conventional storage. Tissue culture or storage at low temperatures is an option for some species. For a variety of reasons many germplasm collections are maintained in fields or gardens, though these collections are vulnerable to natural disasters.

Storage of plant genetic resources as DNA (Adams, 1993) has the advantage of potentially very long lifetimes at relatively low cost. Large numbers of samples can be stored in a relatively small space. Innovations that have been applied to seedbanks, such as robotic access (Figure 2.11; Tsukuba, Japan) and increased security by storing under permafrost (Nordic Genebank), could also be adapted for DNA banks. Recovery of useful genes from DNA samples currently involves technical difficulties related to the storage of seed as a resource for plant improvement. However, increasing knowledge of the structure and arrangement of plant genomes and of gene sequences, together with improvements in the efficiency and general applicability of plant transformation, will increase the ease with which genetic resources can be recovered from a DNA bank and successfully utilized.

The model plant, *Arabidopsis thaliana*, has become an important genetic resource. The small genome size, short generation time and small size of the plants has made this species an ideal plant to use for gene analysis and cloning. An international network of *Arabidopsis* resource centres has been established (Scholl and Anderson, 1994).

Molecular analysis of genetic diversity in plant genetic resource collections allows better management, especially when resources and space are limited as in most collections (Marshall and Brown, 1975). Duplicate accessions may be identified and eliminated and accessions that contain large numbers of unique genes may be able to be targeted for special protection. Virk *et al.* (1995a) evaluated RAPD analysis for the identification of duplicates in a rice germplasm collection. The analysis of 100 markers was suggested as a practical criterion for designating two samples as duplicates. Samples might then be discarded if no morphological differences were detectable. Only sequencing of the entire genomes would guarantee that the samples were identical, but some limit is required for operational purposes. RAPD analysis has been applied to the analysis of the rice germplasm collection held at the International Rice Research Institute (Virk *et al.*, 1995b)

Figure 2.12 Analysis of genetic variation in plant genetic resource collections. A collection of wild barley (*Hordeum spontaneum*) was analysed to show the extent to which genetic variation in specific geographic regions was represented (Weining and Henry, 1995). Some of the accessions from Iran in this collection showed little or no genetic differences while the material from Israel represented in the collection was relatively diverse. This may reflect either the extent of variation of the genotypes in these regions or the adequacy of sampling of material from different populations for inclusion in this collection.

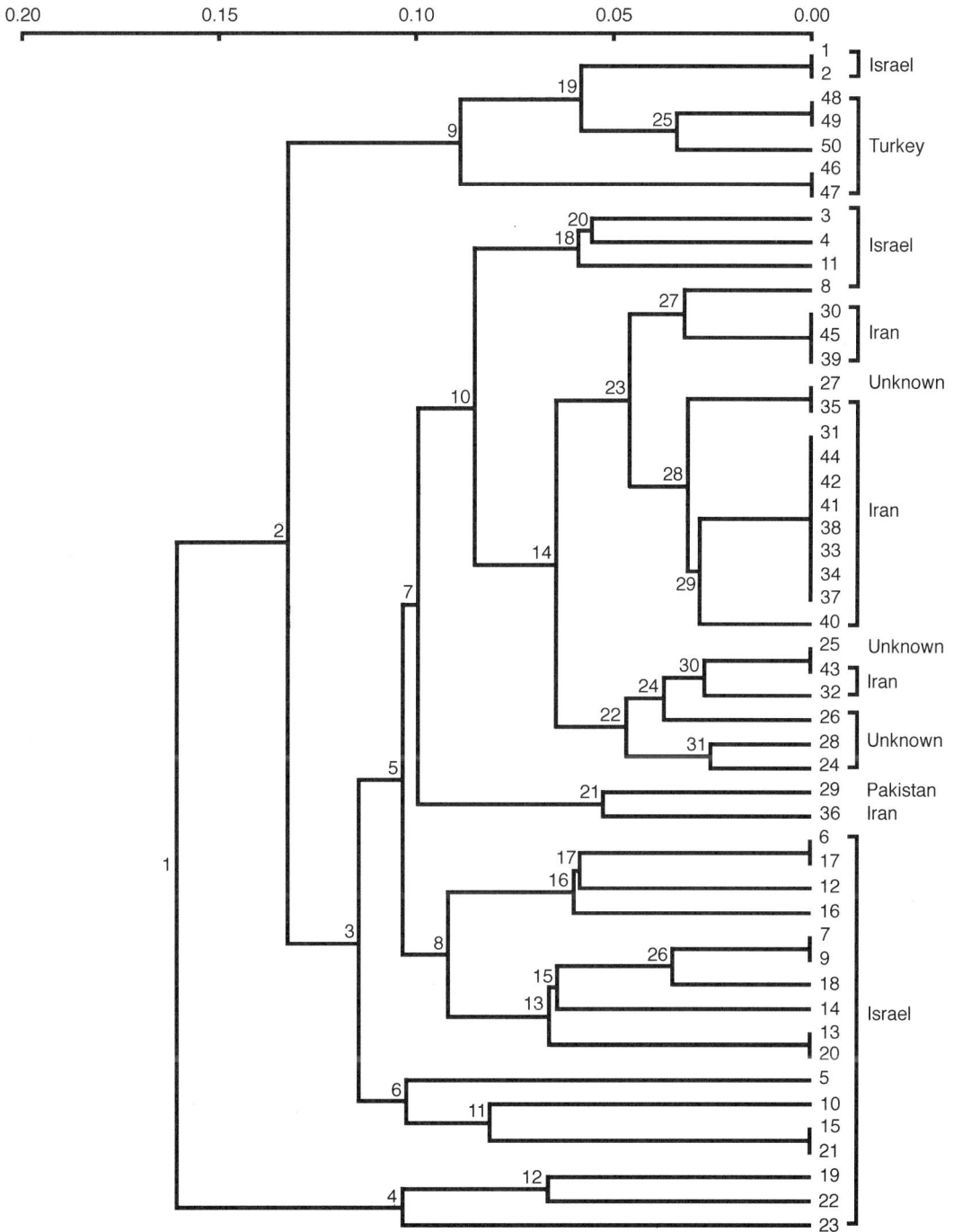

Exploration

↓

Classification ————————————————→ Conservation (*in situ*)

↓

Collection ————————————————————→ **Conservation (*ex situ*)**

↓

Evaluation

↓

Utilization (plant breeding)

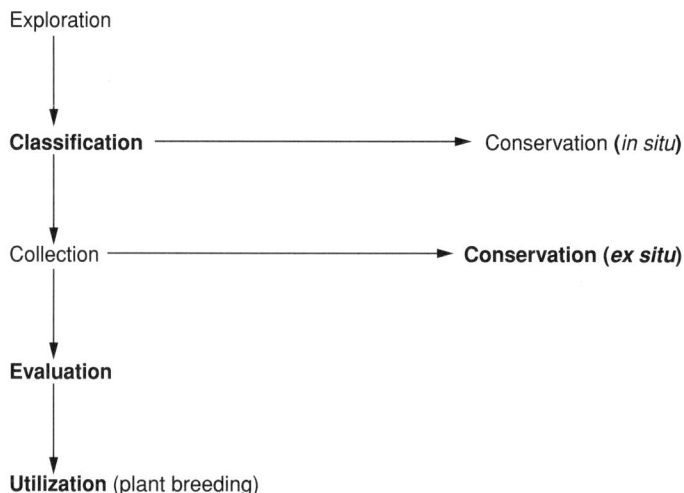

Figure 2.13 Steps in the identification and utilization of plant genetic resources where molecular analysis can be applied. Steps in which molecular methods may be used are shown in bold.

Molecular analysis may suggest the need for further collections from particular locations or of specific genetic types. RFLP analysis of onion (*Allium cepa*) germplasm has shown that short day onions, while not distinct from long day onions, were more genetically diverse indicating the value in including short day onions in germplasm collections (Bark and Havey, 1995). Molecular analysis of wild barley accessions held in an Australian collection was used to identify differing level of polymorphism in lines from different sources (Weining and Henry, 1995; Figure 2.12). Lawrence *et al.* (1995) suggest that about 172 plants collected at random will sample almost all genes present at a frequency of at least 5%. Molecular analysis can be used to verify or refine this type of guideline for specific plant populations.

Molecular markers may be especially useful in species that are difficult to maintain in germplasm collections. RAPD analysis had been applied to the assessment of *Musa* (banana) germplasm (Bhat and Jarrret, 1995) which is usually maintained in the field or in tissue culture. Molecular analysis techniques have application at several stages in the overall process of identification and utilization of plant genetic resources, especially in classification, *ex situ* conservation, evaluation and utilization in plant improvement (Figure 2.13). Molecular analysis may even be useful in identification of the most effective method for the conservation of the genetic resources, allowing comparison of the likely outcomes of efforts at conservation *in situ* (in the wild) or *ex situ* (in germplasm collections).

2.6.3 APPLICATIONS IN PLANT IMPROVEMENT

Plant breeders may use molecular analysis of their genetic resource collections in the selection of parents for use in crosses. For example, molecular analysis can be used to identify the most distantly related individuals in a collection to allow the widest possible crosses in an attempt to gain maximum 'hybrid vigour' or heterosis advantage (Chapter 3). In other cases, the breeder may wish to introduce a new trait controlled by a single gene or a small number of genes, without altering the general genetic background contributing to many favourable quantitative traits. This could be attempted by using the most closely related individual with the new trait.

Molecular markers can be used to determine relationships between breeding lines and to compare pedigree-based assessment of relationships. A study of European barley germplasm suggests that RFLP data may be a better guide to the relatedness of breeding lines than an analysis of ancestry (Graner *et al.*, 1994). The molecular and pedigree-based estimates of relatedness were only poorly correlated. RFLP markers have been used to evaluate the complex relationships between wild and cultivated *Sorghum* subspecies (Cui *et al.*, 1995). Analyses of this type can provide a useful guide to collection of plant genetic resources. Microsatellites have been shown to be useful for the analysis of genetic resources in rapeseed (*Brassica napus*) (Kresowich *et al.*, 1995). RAPD markers have been applied to the analysis of genetic resources in many species. Chapter 3 will detail the use of molecular techniques in plant breeding.

2.7.1 INTRODUCTION

> **2.7 Genetic diversity in plant pathogens**

Understanding genetic variation in plant pathogens may be of great importance in the control of plant diseases. Breeding disease-resistant plants with durable resistance requires that the plants are resistant to all genotypes of the pathogen. Highly variable plant pathogens may prove difficult to combat by development of resistant plant genotypes. Single resistance genes in the plant are likely to result in a rapid selection of pathogen genotypes that can overcome the plant's resistance and rapid spread of these genes in the pathogen population. Genetically uniform pathogens suggest a higher probability of success in developing durable resistance by deployment of single resistance genes.

Gene for gene hypothesis and the interaction between plants and their pathogens

The genetics of plant resistance to disease can be considered in relation to the existence of dominant resistance genes in the plant and a corresponding dominant avirulence gene in the pathogen. The interaction between the plant and the pathogen is determined as indicated below. The absence of either the dominant resistance allele or the avirulence gene results in a compatible (susceptible) interaction. R = dominant resistance gene; A = dominant avirulence gene.

Plant genotype	Pathogen genotype	Plant reaction
R	A	resistant
R	a	susceptible
r	A	susceptible
r	a	susceptible

Quarantine of plant diseases may be better managed with a knowledge of the extent of genetic variation of the pathogen in different regions. Restriction of movement of biological material between sites with different pathogen genotypes should be the objective of quarantine procedures. The study of genetic variation in pathogens may provide clues to the origin of the pathogen and patterns of dispersal as influenced by environmental and human factors. This may suggest other options for the management of the pathogen.

2.7.2 ANALYSIS OF GENETIC DIVERSITY IN PLANT PATHOGENS

Many techniques have been applied to the analysis of genetic variation in plant pathogens. Examples of RFLP, RAPD and other techniques are given below.

(a) RFLP

RFLP analysis of the pathogen causing late blight of potato, *Phytophthora infestans*, has been used to analyse genetic diversity in European populations (Drenth, 1994). The origin of new mating types in the early 1980s was shown to be associated with an increase in genetic variation in the populations.

(b) RAPD analysis

RAPD analysis was used to evaluated genetic variation in 46 isolates of *Fusarium oxysporum* f. sp. *vasinfectum*, the causal organism of wilt in cotton (Assigbetse *et al.*, 1994). Isolates from different geographic areas and different pathogenicity groups were distinguished into clusters by the molecular analysis. UPGMA analysis of the RAPD data from 11 primers identified three major groupings corresponded to the three races distinguished by pathogenicity tests.

(c) Other approaches

Plant pathogens have genomes that differ in size and complexity from those of higher plants. This provides an opportunity to consider approaches that would not be suitable for the analysis of genetic variation in plants. Analysis of the chromosomes by pulsed field gel electrophoresis has been proposed for use in phylogenetic analysis of organisms with chromosomes of appropriate size, such as fungi. This approach was not able to distinguish two isolates of *Phytophthora megasperma* with nine chromosomal DNA bands between 1.4 and 4 Mbp (Howlett, 1990). However, RFLP with ribosomal gene probes was useful for their distinction.

KEY TERMS

Allele
Allozymes
Chloroplast
Cladogram
Dominant
Hardy–Weinberg Equilibrium (HWE)
Heterosis
Heterozygous
Homozygous
Inbreeding
Locus
Mitochondria
Nucleus
Parsimony
Phylogeny
Polymorphism Information Content (PIC)
Unweighted pair group method with arithmetic averaging (UPGMA)

EXAMPLES OF WORKED QUESTIONS

1. Describe some of the consequences of a population being at Hardy–Weinberg equilibrium.

   Allele frequencies do not change from one generation to the next. Rare alleles are much more common in heterozygotes than homozygotes.

2. List factors that cause deviation from Hardy–Weinberg equilibrium.

   Selection, drift, migration and inbreeding.

3. Calculate the dissimilarities, draw up a dissimilarity matrix and generate a phenetic tree by UPGMA analysis from the following marker data taken from an electrophoresis gel in the analysis of three plant samples. Plant X had 75 bands in common with plant Y. Plants X and Z shared 50 bands and plants Y and Z shared 60 bands. All plants had 100 scorable bands.

	X	Y	Z
X	1.00	0.25	0.50
Y	0.25	1.00	0.40
Z	0.50	0.40	1.00

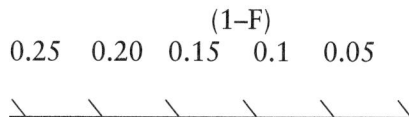

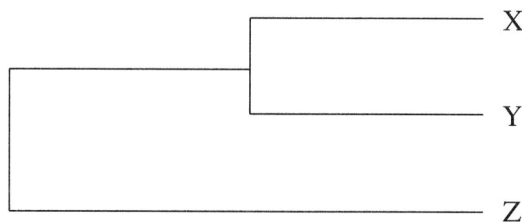

   (1–F)

   0.25   0.20   0.15   0.1   0.05

4. Three samples representing different varieties of wheat are to be distinguished. Which of the following methods could be used?

- RAPD analysis
- SSR analysis
- sequencing of ITS
- chloroplast gene sequencing.

RAPD and SSR could provide sufficient discrimination.

The analysis of genetic variation in plant viruses is possible by comparison of the entire or partial sequence of the viral genome. Comparison of coat protein sequences or the sequences of other major genes is a common option for the analysis of variation in viruses.

Questions

1. Describe the impact of inbreeding on the level of heterozygosity in plant populations.

2. Define the differences between selection, drift and migration and how they may impact on genotype frequencies in a plant population.

3. Analysis of a plant population with alleles A and B present at equal frequencies indicated that 50% of the population was heterozygous at this locus. Was this population outcrossing or predominantly self-pollinating?

4. How would you use molecular analysis to manage a wild population of 120 individuals thought to be the only surviving members of the species?

References

Abo-elwafa, A., Murai, K. and Shimada, T. (1995) Intra- and inter-specific variations in Lens revealed by RAPD markers. *Theoretical and Applied Genetics*, 90, 335–40.

Adams, R.P. (1993) The conservation and utilization of genes from endangered and extinct plants: DNA bank-net, in *Gene Conservation and Exploitation* (eds J.P. Gustafson, R. Appels and P. Raven), Plenum, New York, pp. 35–52.

Anderson, J.A., Churchill, G.A., Autrique, S.D., Tanksley, S.D. and Sorrells, M.E. (1993) Optimizing parental selection for genetic linkage maps. *Genome*, 36, 181–6.

Assigbetse, K.B., Fernandez, D., Dubois, M.P. and Geiger, J.-P. (1994) Differentiation of *Fusarium oxysporum* f. sp. *vasinfectum* races on cotton by random amplified polymorphic DNA (RAPD) analysis. *Phytopathology*, 84, 622–6.

Baldwin, B.G. (1992) Phylogenetic utility of the internal transcribed spacers of nuclear ribosomal DNA in plants: an example from the Compositae. *Molecular Phylogenetics and Evolution*, 1, 3–16.

Bark, O.H. and Havey, M.J. (1995) Similarities and relationships among populations of the bulb onion as estimated by nuclear RFLPs. *Theoretical and Applied Genetics*, **90**, 407–14.

Barlow, B.A. (1981) The Australian flora: its origin and evolution. *Flora of Australia*, **1**, 25–75

Baverstock, P.R. and Moritz, C. (1990) Sampling design, in *Molecular Systematics* (eds D.M. Hillis and C. Moritz), Sinauer Associates, Sunderland, pp. 13–24.

Bhat, K.V. and Jarret, R. L. (1995) Random amplified polymorphic DNA and genetic diversity in Indian *Musa* germplasm. *Genetic Resources and Crop Evolution*, **42**, 107–18.

Bold, H.C. (1961) *The Plant Kingdom*, Prentice-Hall Inc., Englewood Cliffs.

Bolger, D.J. and Simpson, B.B. (1995) A chloroplast DNA study of the Agavaceae. *Systematic Botany*, **20**, 191–205.

Bradshaw, H.D. Jr, Wilbert, S.M., Otto, K.G. and Schemske, D. W. (1995) Genetic mapping of floral traits associated with reproductive isolation in monkeyflowers (Milmulus). *Nature*, **376**, 762–5.

Campbell, N.J.H., Harriss, F.C., Elphinstone, M.S. and Baverstock, P.R. (1995) Outgroup heteroduplex analysis using temperature gradient gel electrophoresis: high resolution, large scale, screening of DNA variation in mitochondrial control region. *Molecular Ecology*, **4**, 215–26.

Chin, H.F. (1994) Seedbanks: conserving the past for the future. *Seed Science and Technology*, **22**, 385–400.

Cox, A.V., Bennett, M.D. and Dyer, T.A. (1992) Use of the polymerase chain reaction to detect spacer size heterogeneity in plant 5S-rRNA gene clusters and to locate such clusters in wheat (*Triticum aestivum* L.). *Theoretical and Applied Genetics*, **83**, 684–90.

Cronquist, A. (1981) *An Integrated System of Classification of Flowering Plants*, Colombia University Press, New York.

Crozier, R.H. (1990) From population genetics to phylogeny: uses and limits of mitochondrial DNA. *Australian Systematic Botany*, **3**, 111–24.

Cui, Y.X., Xu, G.W., Magill, C.W., Schertz, K.F. and Hart, G.E. (1995) RFLP-based assay of *Sorghum bicolor* (L.) Moench genetic diversity. *Theoretical and Applied Genetics*, **90**, 787–96.

Cummings, K.W., Szaro, T.M. and Burns, T.D. (1996) Evolution of extreme specialization within a lineage of ectomycorrhizal epiparasites. *Nature*, **379**, 63–6.

Cummings, K.W. (1992) Design and testing of a plant-specific PCR primer for ecological and evolutionary studies. *Molecular Ecology*, **1**, 233–40.

Darbyshire, B. and Henry, R.J. (1979) The association of fructans with high percentage dry weight in onion cultivars suitable for dehydrating. *Journal of the Science of Food and Agriculture*, **30**, 1035–8.

Dean, C. and Schmidt, R. (1995) Plant genomes: a current molecular description. *Annual Review of Plant Physiology and Plant Molecular Biology*, **46**, 395–418.

de la Cruz, M., Gomez-Pompa, R.W.A. and Mota-Bravo, L. (1995) Origins of cacao cultivation. *Nature*, **375**, 452–3.

Demeke, T. and Adams, R.P. (1994) The use of PCR–RAPD analysis in plant taxonomy and evolution, in *PCR Technology Current Innovations* (eds

H.G. Griffin and A.M. Griffin), CRC Press, Boca Raton, pp. 179–200.

Doyle, J.J. and Doyle, J.L. (1991) DNA and higher plant systematics: some examples from the legumes, in *Molecular Techniques in Taxonomy* (eds G.M. Hewitt, A.W.B. Johnston and J.P.W. Young), Springer-Verlag, Berlin, pp. 101–15.

Drenth, A. (1994) *Molecular genetic evidence for a new sexually reproducing population of* Phytophthora infestans *in Europe*. PhD Thesis, University of Wageningen, The Netherlands.

FAO Yearbook (1991) Food and Agriculture Organization of the United Nations, vol. 45, p. 61.

Graham, G.C., Henry, R.J., Godwin, I.D. and Nikles, D.G. (1996) Phylogenetic position and evolution of *Araucaria cunninghamii* based upon 18S rRNA gene sequences. *Australian Journal of Systematic Botany* (in press)

Graner, A., Ludwig, W.F. and Melchinger, A.E. (1994) Relationships among European barley germplasm: II. Comparison of RFLP and pedigree data. *Crop Science*, **34**, 1199–205.

Heywood, V.H. (1978) *Flowering Plants of the World*, Oxford Press, London.

Hillis, D.M. and Dixon, M.T. (1991) Ribosomal DNA: Molecular evolution and phylogenetic inference. *The Quarterly Review of Biology*, **66**, 411–53.

Hirai, A. and Nakazono, M. (1993) Six percent of the mitochondrial genome of rice came from chloroplast DNA. *Plant Molecular Biology Reporter*, **11**, 98–100.

Hosaka, K. (1995) Successive domestication and evolution of the Andean potatoes as revealed by chloroplast DNA restriction endonuclease analysis. *Theoretical and Applied Genetics*, **90**, 356–63.

Howlett, B.J. (1990) Pulsed field gel electrophoresis as a method for examining relationships between organisms; its application to the genus *Phytophthora. Australian Systematic Botany*, **3**, 75–80.

Hsiao, C., Chatterton, N.J., Asay, K.H. and Jensen, K.B. (1995) Molecular phylogeny of the Pooideae (Poaceae) based on nuclear rDNA (ITS) sequences. *Theoretical and Applied Genetics*, **90**, 389–98.

Huff, D.R., Peakall, R. and Smouse, P.E. (1993) RAPD variation within and among natural populations of outcrossing buffalograss (*Buchloe dactyloides* (Nutt.) Engelm.). *Theoretical and Applied Genetics*, **86**, 927–34.

Ishii, T., Terachi, T., Mori, N. and Tsunewaki, K. (1993) Comparative study on the chloroplast, mitochondrial and nuclear genome differentiation in two cultivated rice species, *Oryza sativa* and *O. glaberrima*, by RFLP analyses. *Theoretical and Applied Genetics*, **86**, 88–96.

Kanis, A. (1981) An introduction to the system of classification used in the flora of Australia, in *Flora of Australia*, Australia Government Publishing Service, Canberra, Vol. 1, pp. 77–111.

Kawata, M. Ohmiya, A., Shimamoto, Y., Oono, K. and Takaiwa, F. (1995) Structural changes in the plastid DNA of rice (*Oryza sativa* L.) during tissue culture. *Theoretical and Applied Genetics*, **90**, 364–71.

Kimura, M. (1980) A simple method for estimating evolutionary rates of base substitutions through comparative studies of nucleotide sequences. *Journal of Molecular Evolution*, **16**, 111–20.

Kresovich, S., Szewc-McFadden, A.K., Bliek, S.M. and McFerson, J.R. (1995) Abundance and characterization of simple-sequence repeats (SSRs) isolated from a size-fractionated genomic library of *Brassica napus* L. (rapeseed). *Theoretical and Applied Genetics*, **91**, 206–11.

Lamboy, W.F. (1994) Computing genetic similarity coefficients from RAPD data: correcting for the effects of PCR artefacts caused by variation in experimental conditions. *PCR Methods and Applications*, **4**, 38–43.

Lawrence, M.J., Marshall, D.F. and Davies, P. (1995) Genetics of genetic conservation. I. Sample size when collecting germplasm. *Euphytica*, **84**, 89–99.

Leigh, J., Boden, R. and Briggs, J. (1994) *Extinct and Endangered Plants of Australia*, Macmillan, Melbourne.

Luo, H., Van Coppenolle, B., Sequin, M. and Boutry, M. (1995) Mitochondrial DNA polymorphism and phylogenetic relationships in *Hevea brasiliensis*. *Molecular Breeding*, **1**, 51–63.

Lynch, A.J.J. and Vaillancourt, R.E. (1995) Genetic diversity in the endangered *Phebalium daviesii* (Rutaceae) compared to that in two widespread congeners. *Australian Journal of Botany*, **43**, 181–91.

Lynch, M. and Milligan, B.G. (1994) Analysis of population genetic structure with RAPD markers. *Molecular Ecology*, **3**, 91–9.

Marshall, D.R. and Brown, A.H.D. (1975) Optimum sampling strategies in genetic conservation, in *Crop Genetic Resources for Today and Tomorrow* (eds O.H. Frankel and J.G. Hawkes), Cambridge University Press, London, pp. 53–80.

May, C.E. and Appels, R. (1987) Variability and genetics of spacer DNA sequences between the ribosomal-RNA genes of hexaploid wheat (*Triticum aestivum*). *Theoretical and Applied Genetics*, **74**, 617–24.

Muza, F.R., Lee, D.J., Andrews, D.J. and Gupta, S.C. (1995) Mitochondrial DNA variation in finger millet (*Eleusine coracana* L. Gaertn). *Euphytica*, **81**, 199–205.

Nei, M. and Li, W.H. (1979) Mathematical model for studying genetic variation in terms of restriction endonucleases. *Proceedings of the National Academy of Science, USA*, **74**, 5269–73.

N'Goran, J.A.K., Laurent, V., Risterucci, A.M. and Lanaud, C. (1994) Comparative genetic diversity studies of *Theobroma cacao* L. using RFLP and RAPD markers. *Heredity*, **73**, 589–97.

Peakall, R., Smouse, P.E. and Huffs, D.R. (1995) Evolutionary implications of allozyme and RAPD variation in diploid populations of dioecious buffalograss *Buchloe dactyloides*. *Molecular Ecology*, **4**, 135–47.

Peever, T.L. and Milgroom, M.G. (1994) Genetic structure of *Pyrenophora teres* populations determined with random amplified polymorphic DNA markers. *Canadian Journal of Botany*, **72**, 915–23.

Pimm, S.L., Russell, G.J., Gittleman, J.L. and Brooks, T.M. (1995) The future of biodiversity. *Science*, **269**, 347–9.

Raina, S.N. and Ogihara, Y. (1995) Ribosomal DNA repeat unit polymorphism in 49 *Vicia* species. *Theoretical and Applied Genetics*, **90**, 477–86.

Rajora, O.P. and Dancik, B.P. (1995a) Chloroplast DNA variation in *Populus*. I. Intraspecific restriction fragment diversity within *Populus deltoides*, *P. nigra* and *P. maximowiczii*. *Theoretical and Applied Genetics*, **90**, 317–23.

Rajora, O.P. and Dancik, B.P. (1995b) Chloroplast DNA variation in *Populus*. II. Interspecific restriction fragment polymorphisms and genetic relationships among *Populus deltoides*, *P. maximowiczii.*, and *P. × canadensis*. *Theoretical and Applied Genetics*, **90**, 324–30.

Rajora, O.P. and Dancik, B.P. (1995c) Chloroplast DNA variation in *Populus*. III. Novel chloroplast DNA variants in natural *Populus × canadensis* hybrids. *Theoretical and Applied Genetics*, **90**, 331–4.

Ritland, C.E., Ritland, K. and Straus, N.A. (1993) Variation in the ribosomal internal transcribed spacers (ITS1 and ITS2) among eight taxa of the *Mimulus guttatus* species complex. *Molecular Biology and Evolution*, **10**, 1273–88.

Roder, M., Plaschke, J., Konig, S. U., Borner, A., Sorrells, M.E., Tanksley, S.D. and Ganal, M.W. (1995) Abundance, variability and chromosomal location of microsatellites in wheat. *Molecular and General Genetics*, **246**, 327–33.

Russell, J.R., Hosein, F., Johnson, E., Waugh, R. and Powell, W. (1993) Genetic differentiation of cocoa (*Theobroma cacao* L.) populations revealed by RAPD analysis. *Molecular Ecology*, **2**, 89–97.

Savard, L., Michaud, M. and Bousquet, J. (1993) Genetic diversity and phylogenetic relationships between birches and alders using ITS, 18S rRNA, and *rbc*L gene sequences. *Molecular Phylogenetics and Evolution*, **2**, 112–18.

Scholl, R. and Anderson, M. (1994) Arabidopsis biological resource center. *Plant Molecular Biology Reporter*, **12**, 242–4.

Solignac, M., Pelandakis, M., Rousset, F. and Chenvil, A. (1991) Ribosomal RNA phylogenies, in *Molecular Techniques in Taxonomy* (eds G.M. Hewitt, A.W.B. Johnston and J.P.W. Young), Springer-Verlag, Berlin, pp. 73–85.

Sun, Y., Skinner, D.Z., Liang, G.H. and Hulbert, S.H. (1994) Phylogenetic analysis of sorghum and related taxa using internal transcribed spacers of nuclear ribosomal DNA. *Theoretical and Applied Genetics*, **89**, 26–32.

Szmidt, A.E. (1994) Molecular population genetics and evolution: two missing elements in studies of biodiversity, in *Measuring and Monitoring Biodiversity in Tropical and Temperate Forests* (eds T.J.B. Boyle and B. Boontawee), Centre for International Forestry Research, Bogor, pp. 177–93.

Taylor, M.F.J., Shen, Y. and Kreitma, M.E. (1995) A population genetic test of selection at the molecular level. *Science*, **270**, 1497–9.

Tilman, D., Wedin, D. and Knopps, J. (1996) Productivity and sustainability influenced by biodiversity in grassland ecosystems. *Nature*, **379**, 718–20.

Virk, P.S., Newbury, H.J., Jackson, M.T. and Ford-Lloyd, B.V. (1995a) The identification of duplicate accessions within a rice germplasm collection using RAPD analysis. *Theoretical and Applied Genetics*, **90**, 1049–55.

Virk, P.S., Ford-Lloyd, B.V., Jackson, M.T. and Newbury, H.J. (1995b) Use of RAPD for the study of diversity within plant germplasm collections. *Heredity*, **74**, 170–9.

Wakasugi, T., Tsudzuki, J., Ito, S., Shibata, M. and Sugiura, M. (1994) A physical map and clone bank of the black pine (*Pinus thunbergii*) chloroplast genome. *Plant Molecular Biology Reporter*, **12**, 227–41.

Warwick, S.I. and Black, L.D. (1994) Evaluation of the subtribes Moricandinae, Savignyinae, Vellinae, and Zillinae (Brassicaceae, tribe Brassiceae) using chloroplast DNA restriction site variation. *Canadian Journal of Botany*, **72**, 1692–701.

Weining, S. and Henry, R.J. (1995) Molecular analysis of the DNA polymorphism of wild barley (*Hordeum spontaneum*) germplasm using the polymerase chain reaction. *Genetic Resources and Crop Evolution*, **42**, 273–81.

West, J.G. and Faith, D.P. (1990) Data, methods and assumptions in phylogenetic inference. *Australian Journal of Systematic Botany*, **3**, 9–20.

Yu, K. and Pauls, K.P. (1994) The use of RAPD analysis to tag genes and determine relatedness in heterogeneous plant populations using tetraploid alfalfa as an example, in *PCR Technology Current Innovations* (eds H.G. Griffin and A.M. Griffin), CRC Press, Boca Raton, pp. 201–14.

Zimmer, E.A., Hamby, R.K., Arnold, M.L., Leblanc, D.A. and Theriot. E.C. (1989) Ribosomal RNA phylogenies and flowering plant evolution, in *The Hierarchy of Life* (eds B. Fernholm, K. Bremer and H. Jornvall), Elsevier Science Publishers, pp. 205–14.

Zurawski, G. and Clegg, M.T. (1987) Evolution of higher-plant chloroplast DNA-encoded genes: Implications for structure–function and phylogenetic studies. *Annual Review of Plant Physiology*, **38**, 391–418.

Molecular markers in plant improvement

3

Wheat breeding field plots. Plant breeding may involve the evaluation of large numbers of genotypes. Molecular markers are a useful tool for use in assessing lines in plant breeding.

Chapter outline

Plant breeding is essential for the maintenance of world food supply. The growth of population (Figure 3.1) requires the production of more and more food from an environment with a reducing capacity to support plant growth. Changes in land use patterns (Table 3.1) do not provide a basis for meeting the increase in demand for food and other products derived from plants. In 1995, it was estimated that

3.1 Introduction

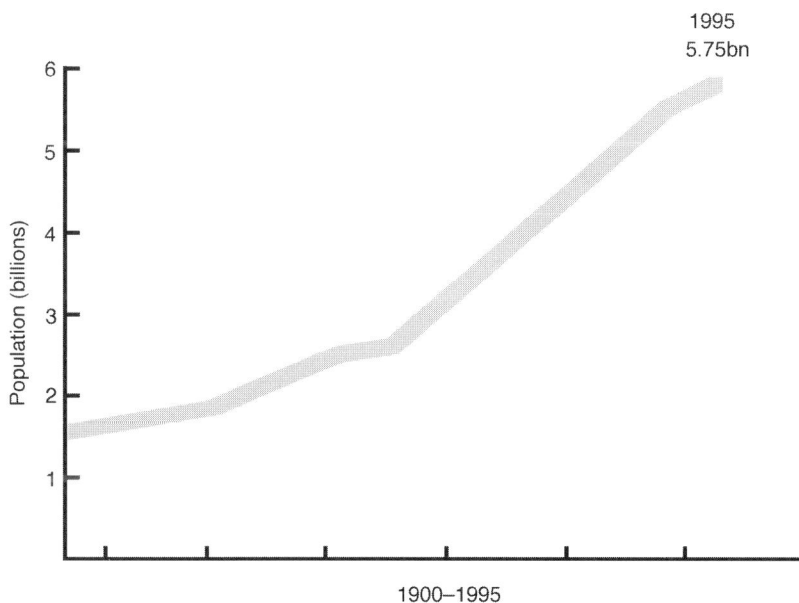

Figure 3.1 World population growth. The increasing demand for food, fibre and forest resources requires the careful application of molecular techniques to plant improvement.

Table 3.1 World land use by productivity class with trends over 25 years to 2000. (From Hall and Scurlock, 1993.)

	Land use (1975) (M ha)	Land use (2000) (M ha)	Net change (M ha)
Cropland			
High	400	345	−55
Medium	500	745	+245
Low	600	710	+110
Total	1500	1800	+300
Grassland			
High	200	170	−30
Medium	300	320	+20
Low	500	510	+10
Zero	2000	2000	0
Total	3000	3000	0
Forest			
High	100	30	−70
Medium	300	100	−200
Low	400	230	−170
Zero	3300	3140	−160
Total	4100	3500	−600
Non-agricultural	400	600	+200
Other land	4400	4500	+100

nearly one-third of the world's arable land had been lost in the preceding 40 years and an additional quarter of a million people required food each day (Pimentel *et al.*, 1995). Genetic improvement of plants is the only option available to satisfy this demand. Very large quantities of food are now produced from relatively few species (Table 2.4). Genetic improvement of these species may be viewed as crucial for human survival. Forest and fibre resources are also stretched by population growth, creating an urgent need for genetic improvement of these plant species.

Genetic engineering of plants (Chapter 4) offers solutions to some of these problems, but the most immediate gains from the application of molecular biology are probably to be found in the use of molecular markers in plant breeding. This chapter will outline the available techniques and their practical application in plant improvement.

3.2.1 INTRODUCTION

The molecular marker techniques available for use in plant improve-ment have been introduced in Chapter 1. Their application in plant breeding may require adaptation because of the large numbers of samples to be analysed or because of the small samples available in the early stages of a breeding programme. Sample size restrictions have been largely eliminated by the use of PCR-based methods with great sensitivity. Specific modifications of PCR protocols may also be devel-oped for use in plant breeding (Sorbral and Honeycutt, 1993). Automation of sample handling and DNA extraction and analysis may be necessary for some applications. Cost may determine the choice of molecular marker technique in many plant breeding applications (Ragot and Hoisington, 1993).

> **3.2 Molecular marker techniques and their application in plant improvement**

3.2.2 EXAMPLES OF APPLYING MOLECULAR MARKERS

Molecular markers may be used for plant identification (Chapter 1) and for analysis of genetic variation in germplasm available for plant improvement (Chapter 2). However, marker-assisted selection is the most common objective in applying molecular markers to plant breeding. Other practical applications are also possible (Poulsen *et al.*, 1996). Several examples, including identification of seed lots, confir-mation of pollination, and assessment of the genetic purity of breeding lines, will be described here. These practical applications may be over-looked by plant breeders focusing only on the use of marker assisted selection for desirable traits.

(a) Confirming the identity of breeding lines

Mislabelling is a common problem in plant breeding. The handling of large numbers of lines provides opportunities for the confusion of sample identity. Molecular markers can be used to confirm suspected mislabelling.

(b) Establishing hybrid identity

The hybrid nature of individuals or seed lots can be established using markers. This may be especially useful in species in which self-polli-nation is a possibility. Somatic hybrids may be identified easily using RAPD analysis (Xu *et al.*, 1993).

(c) Testing for purity of breeding lines

Contamination of plant breeding lines can result from accidental mixing of seed samples, cross-contamination in seed harvesters, and plants arising from seed from earlier crops on the same site. Molecular markers may be used to establish the purity and level of contamination in breeding lines.

(d) Evaluation of somaclonal variation

Somaclonal variation (Larkin and Scowcroft, 1981) may be generated deliberately by plant breeders or may arise as an unwanted consequence of propagation or genetic transformation protocols. Molecular markers allow an assessment of the extent of somaclonal variation. The verification of minimal genetic change is important in attempts to introduce single genes into an otherwise ideal genotype by transformation. If the level of somaclonal variation is too high, transgenic plants provide useful parents for plant breeders rather than genotypes ready for commercial release. In some cases routine backcrossing of the transformal plant to the untransformed genotype may be required to reduce the risk of unwanted somaclonal changes.

(e) Predicting hybrid performance

The enhanced performance of hybrid crops (heterosis) may be related to the genetic distance between the parents used in the cross. Molecular markers can be used to estimate genetic distances between possible parents (Chapter 2) and this information may be used to attempt a prediction of the crosses most likely to perform well as a result of heterosis. Analysis of hybrid rice using RFLP and SSR markers suggests that specific markers may be chosen that are much more useful than randomly selected markers for these predictions (Zhang *et al.*, 1995).

(f) Identifying germplasm likely to contribute useful traits

Molecular markers may be used to evaluate genetic resources and identify possible parents for use in breeding. A survey of rice germplasm using RAPD markers (Virk *et al.*, 1996) showed linkage between the presence of specific markers and QTLs for important traits. A single marker could explain about 50% of the variation in culm number.

3.3.1 INTRODUCTION

Marker-assisted selection offers great opportunity for improved efficiency and effectiveness in selection of plant genotypes with the desired combinations of traits. This approach relies upon the establishment of a linkage between a molecular marker and the characteristic to be selected. Once this has been achieved, breeding selection can be conducted in the laboratory and does not require the expression of the associated phenotype (e.g. disease resistance can be evaluated in the absence of the disease and adult plant characters such as resistance to stress can be assessed in seedlings).

> **3.3 Marker-assisted selection in a plant breeding programme**

3.3.2 MOLECULAR MARKER MAPS OF PLANT GENOMES

Several approaches are available for identifying markers linked to traits of interest. Molecular marker maps are now available for most important species (O'Brien, 1993; Schwarzacher, 1994). Important crop species such as rice have been the subject of intensive mapping efforts (Kurata *et al.*, 1994). Detailed maps have been generated for the human genome using microsatellites (Dib *et al.*, 1996) and similar technology is being applied to plants.

The model plant, *Arabidopsis thaliana*, has been extensively studied. *Arabidopsis* has several advantages as a model system, a very small genome, a short generation time, a small plant size (allowing many plants to be grown in a small space), and large collections of mutants are available for genetic studies. Mapping of plant genomes is a major collaborative process with newsletters, coordination groups and databases. The numbers of markers on plant gene maps may exceed 1000 but the markers are not likely to be evenly distributed over the chromosomes with markers clustered in some regions and absent in others. The frequency of recombination varies at different places with 'hot' and 'cold' spots of recombination being distributed along the length of chromosomes (Schmidt *et al.*, 1995). The relationship between genetic and physical maps suggests that recombination is suppressed near centromeres.

The large size of many plant genomes is shown in Table 3.2. The identification of molecular markers close to useful genes may be easier in smaller genomes. Mapping of genes in a smaller but related genome is an important strategy for location of genes in complex genomes. For example, rice is a useful starting point for the mapping of a gene in a more complex grass species such as wheat (Figure 3.2). The analysis of cereal genomes is assisted by the close relationships between these important crop species (Moore *et al.*, 1993). Polymorphic markers from related species are good sources of markers (e.g. RFLP) for

Table 3.2 Sizes of plant genomes. (From Croy, 1993.)

Species	Ploidy	Genome size (bp) (1 C value)
Allium cepa L.	2	1.72×10^{10}
Anemone blanda	2	1.31×10^{10}
Antirrhinum majus L.	2	1.54×10^{9}
Arabidopsis thaliana	2	1.9×10^{8}
Arachis hypogeae L.	4	1.7×10^{9}
Avana sativa L.	6	1.3×10^{10}
Brassica campestris	2	7.7×10^{8}
Brassica napus L.	2	1.5×10^{9}
Brassica oleracea	2	8.7×10^{8}
Capicum annuum L.	2	5.2×10^{9}
Datura innoxia Miller	2	2.2×10^{9}
Lotus corniculatus	4	0.96×10^{9}
Lupinus albus	2	5.8×10^{8}
Lycopersicon esculentum Miller	2	2.2×10^{9}
Nicotiana tabacum L.	4	3.7×10^{9}
Oryza sativa L.	2	5.8×10^{8}
Petunia hybrida	2	1.5×10^{9}
Phaseolus vulgaris L.	2	1.7×10^{9}
Pisum sativum L.	2	5.0×10^{9}
Secale cereale L.	2	8.5×10^{9}
Solanum tuberosum L.	4	2.0×10^{9}
Sorghum caudatum	4	4.7×10^{9}
Triticum aestivum L.	6	1.7×10^{10}
Triticum monococcum L.	2	5.9×10^{9}
Vicia faba L.	2	1.2×10^{10}
Zea mays L.	2	3.2×10^{9}

constructing maps. The high degree of synteny in the genomes of many important plant species makes study of all plants of at least some potential practical value. Remarkably, QTLs for traits such as grain size that were the subject of independent selection during the domestication of the cereals (sorghum, rice and maize) correspond in their map locations (Paterson *et al.*, 1995). This indicates that synteny may be a powerful tool for the analysis of QTLs in plants. However, it is important to note that repetitive sequences do not share the conservation found in coding genes. RFLP loci (often detected with cDNA probes) can be mapped in related taxa at high frequency but microsatellites are usually unique to each species.

The choice of parents and crosses for use in constructing maps is crucial. Wide crosses may provide more polymorphisms but these polymorphisms may not be found in breeders' crosses, making the map

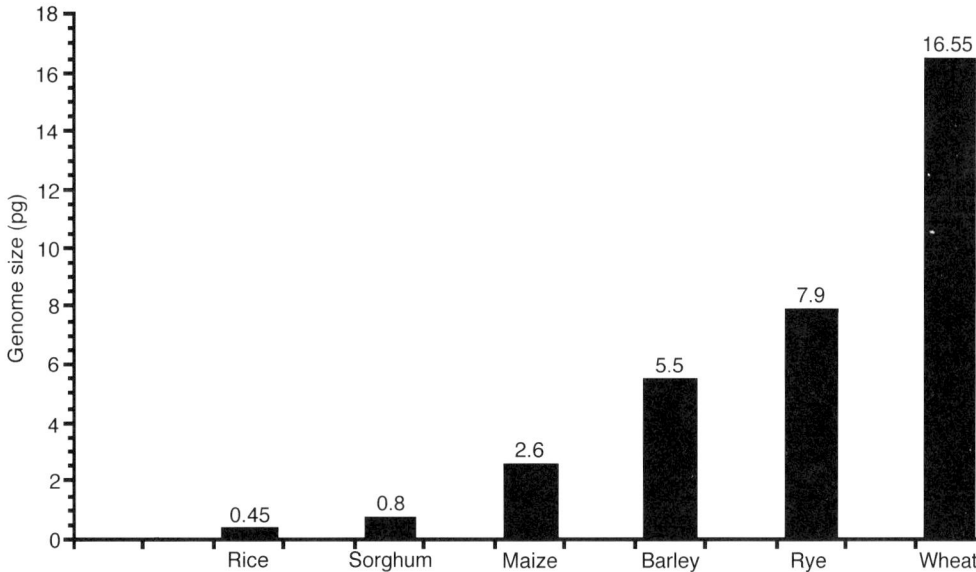

Figure 3.2 Sizes of genomes in the cereals (Poaceae) Note the small size of the rice genome relative to wheat. Rice is thus an easier target for gene mapping and isolation with results being transferable to the larger but closely related genomes of the other major cereals.

unusable in plant breeding. The size of mapping populations will determine the ultimate resolution of the map, less than 50 individuals is unlikely to be useful for generating a map and high-resolution mapping or targeting of specific regions of the genome may require populations of 1000 or more (Young, 1994).

Software such as Mapmaker is available for linkage analysis. Mapmaker performs multipoint linkage analysis. A LOD score is calculated as the log (to base 10) of the ratio between the odds of the hypothesis that a linkage exists and the odds that a linkage does not exist (Huhn, 1995). Some maps have loci that are not linked (Berry *et al.*, 1995) or numbers of linkage blocks that exceed the known number of chromosomes.

3.3.3 LINKAGE OF MOLECULAR MARKERS TO USEFUL TRAITS

The position of genes determining useful traits may be established by analysis of linkage with markers on an established map of the plant's genome. Analysis of any polymorphic markers and any segregating gene (trait) can be used to identify linked markers. Establishment of a linkage to a mapped marker has the advantage that other linked

Comparative mapping

The comparison of genetic maps of related species is illustrated for a hypothetical case below. Species A and B show the same order of markers. Species C has a region of the chromosome (markers 1 to 4) inverted relative to species A and B.

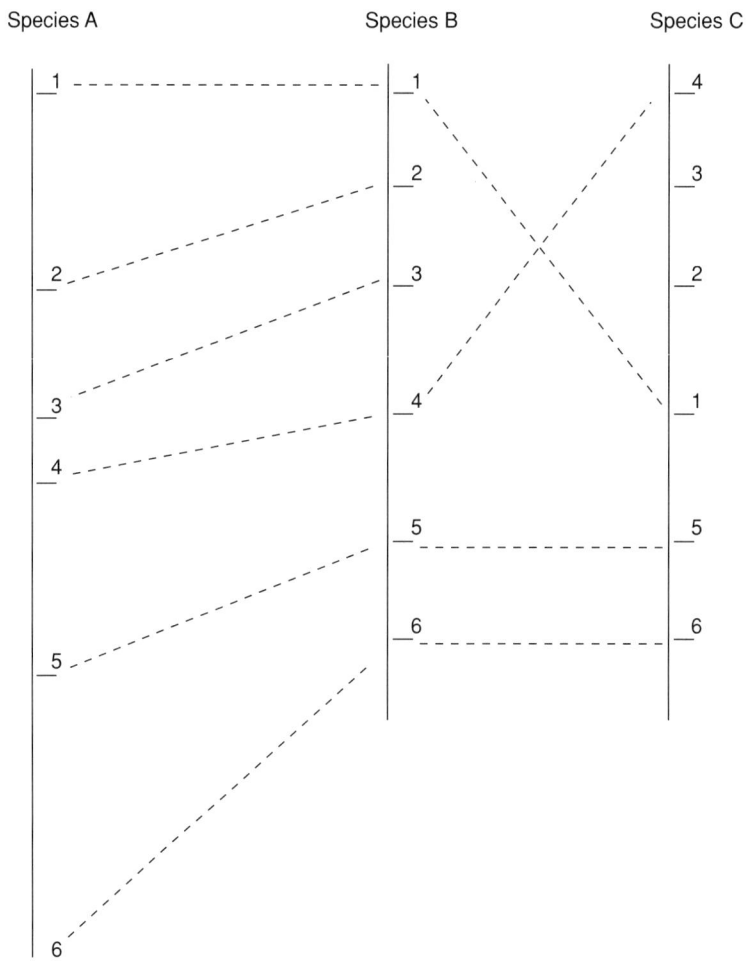

markers from the map are available for use in genetic backgrounds that lack the marker. However, it is possible to find a linked marker that has not been mapped by directly searching for linked polymorphisms in unmapped markers. Bulked segregant analysis (BSA) is a

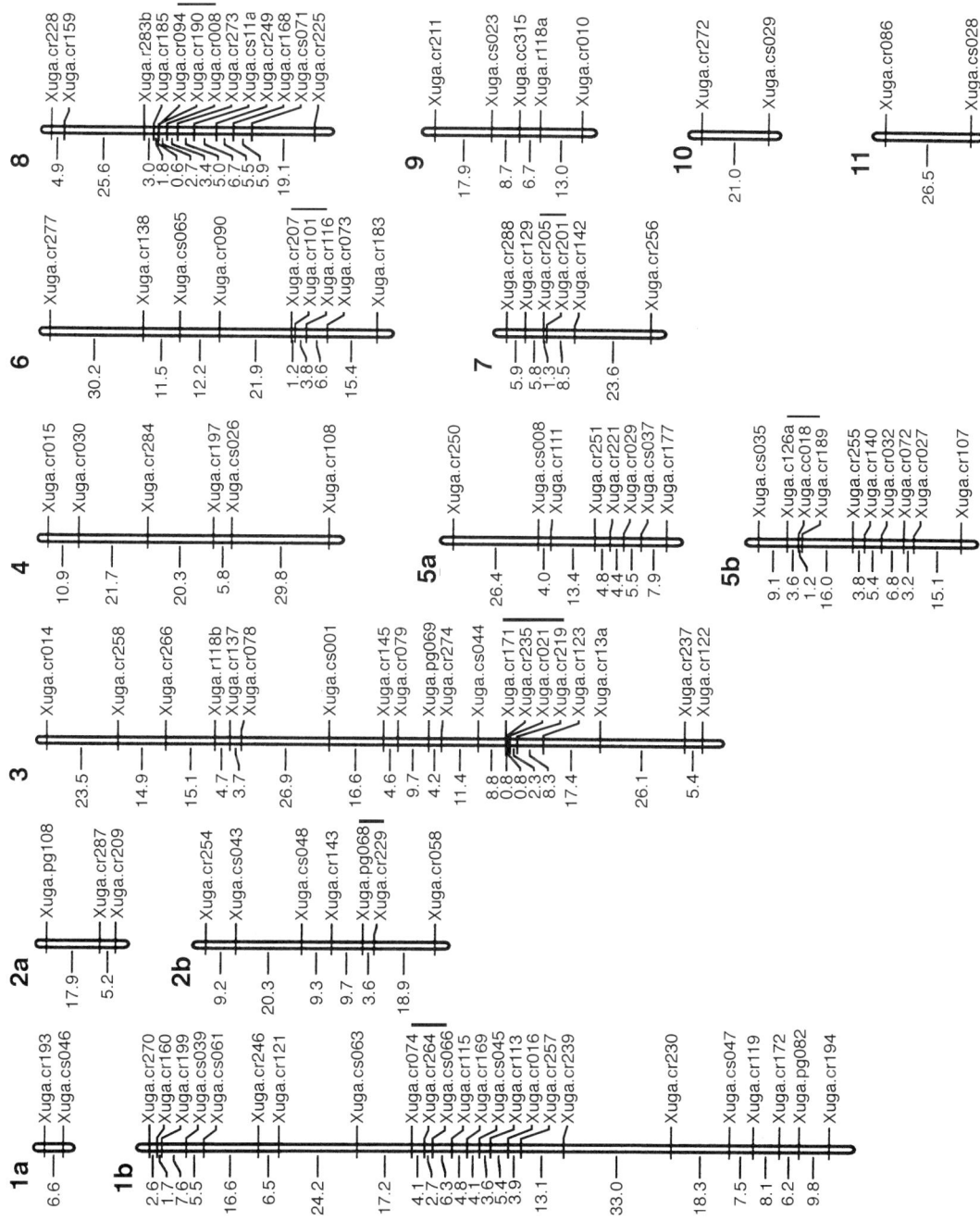

Figure 3.3 Peanut RFLP map.

RFLP map of peanut

Cultivated peanut (*Arachis hypogaea*) is an allotetraploid, apparently of relatively recent origin, displaying very little DNA polymorphism. Generation of a map has required the use of a cross between two closely related wild diploid species. The map shown in Figure 3.3 was produced for a cross (*Arachis stenosperma* × *Arachis cardenasii*) by Halward *et al.* (1993).

---

useful technique for quickly finding a linked marker (Michellmore *et al.*, 1991). BSA involves the testing for polymorphism between only two DNA samples made up of bulks of individuals from the segregating population. One bulk contains DNA from individuals with the gene (trait) being targeted, while the other contains DNA from individuals lacking the gene. Sampling from a segregating population results in both of the two bulks containing most genes. However, the trait used to segregate the bulks will be determined by genes differing between the bulks. DNA polymorphisms between the bulks are therefore likely to be linked to genes for the trait. Mutations can be mapped using this approach, with bulks based upon the presence or absence of the mutant phenotype (Williams *et al.*, 1993).

Reliable scoring of the plants for the trait may be critical in determining success in marker identification. More reliable assessment of quantitative traits requires replicated testing. Analysis of quantitative traits in multiple environments becomes essential when the ranking of genotypes for the trait varies with the environment. However, the availability of only one individual of each genotype in a segregating F_2 population of a plant for which vegetative propagation is not practical may prevent replication. Two common approaches have been to generate recombinant inbred lines (RILs) or to produce doubled haploids (DHs). Double haploids are the most reliable because they are totally homozygous but their efficient production is only possible in a relatively small number of species. RILs are a more generally available option. However, heterozygosity may be greater than expected (Paran *et al.*, 1995).

The analysis of linkages for co-dominant (e.g. RFLP or microsatellite) and dominant (e.g. RAPD) markers may require different statistical analysis in mapping with F_2 populations. These differences do not apply to RILs (because almost all loci are homozygous) or to backcross (BC) populations (because the alleles from the recurrent parent are homozygous and from the donor parent heterozygous). Genetic distances (recombination frequencies) can be calculated most accurately for RIL or BC populations using RAPD markers and least accurately for F_2 populations. The multipoint standard deviation is about twice

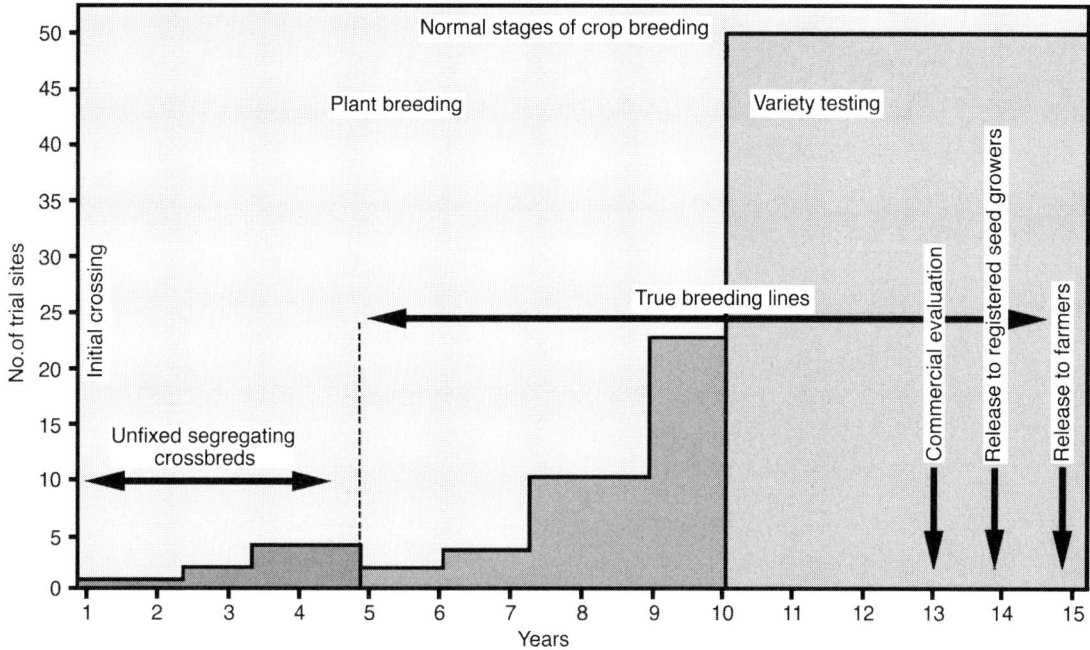

Figure 3.4 Stages in a conventional plant breeding programme. Molecular markers can be used to accelerate this process by providing rapid and reliable assessment of genotype.

as large for a dominant marker than for a co-dominant marker in an F_2 population. DH or other inbred populations allow more efficient mapping (Knapp *et al.*, 1995).

3.3.4 EXAMPLES OF APPLICATION IN A PLANT BREEDING PROGRAMME

Breeding of plants often involves several stages as depicted in Figure 3.4. Crossing of the selected parents may be followed by several rounds of selfing to generate genetically fixed lines. This phase is followed by selection, a process that may require many years of testing, if complex traits subject to environmental influence are under selection. Elite lines may then undergo more extensive testing and commercial evaluation before release. Marker-assisted selection aims to accelerate the whole process, allowing earlier commercial release and also to make each round of selection more efficient.

A plant breeding programme

A PLANT BREEDING PROGRAMME

```
                    ┌─────────────────────┐
                    │ Cross Evaluation Trial │ ◄──────── F3
                    └─────────────────────┘            ▲
                              │                         │
                              │                        F2
                              │                         ▲
                              │                         │
                       ┌──────────────┐                F1
                       │ Selection Trial │               ▲
                       └──────────────┘                 │
                              │                          │
                              │                          │
    Introductions────────► ┌─────────────┐ ──────────► Crossing
                           │ Strain Trial │
                           └─────────────┘
                              │
                              │
                       ┌──────────────┐
                       │ Variety Trial │
                       └──────────────┘
                              │
                              ▼
                        ┌──────────┐
                        │ Release  │
                        └──────────┘
```

In this scheme (Henry *et al.*, 1984) new germplasm is evaluated in a strain trial allowing selection of parents. Crosses are evaluated in a cross evaluation trial before evaluation of larger numbers for the best crosses in a selection trial. Elite material is returned to the strain trials and may progress to variety trials and ultimately commercial release.

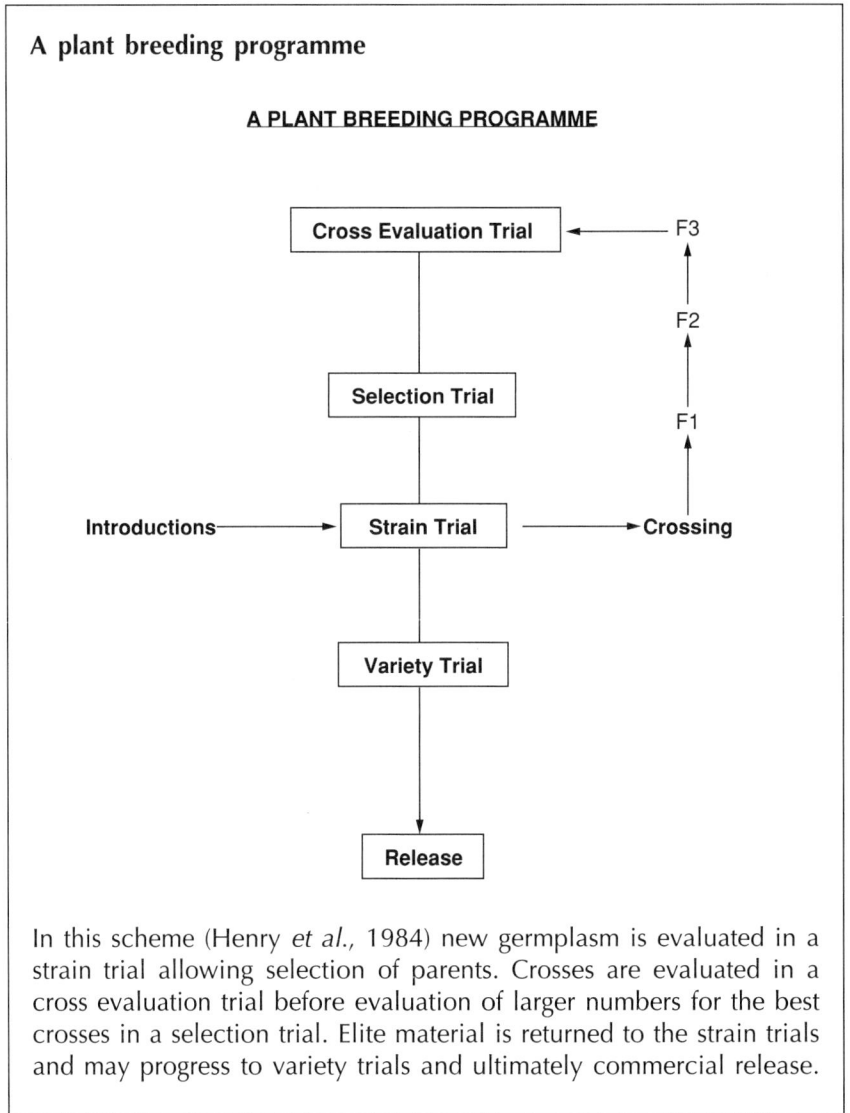

(a) Accelerated backcrossing

Molecular markers may be used to accelerate a backcrossing programme. Plant breeders often need to introduce a single trait (or gene) into a genetic background without altering other characteristics. This is usually achieved by repeated crossing to the plant with the genetic background required. The trait being introduced (e.g. disease resistance) may be selected for at each generation. The number of crosses required varies. Species with more chromosomes, for example, require more generations. Molecular markers allow the selection of individuals with more of the recurrent genome at each generation and thus permit the breeding programme to be completed in fewer generations (Young and Tanksley, 1989).

(b) Selection for a desirable trait

Molecular markers may also be used to directly select for desirable traits. A molecular marker linked to the trait of interest may be screened for at any stage in the breeding program. The strategy to be employed will depend upon various factors. Markers that are more closely linked to the target gene can be applied to fewer generations without significant risk that the required trait will be lost in small populations. Multiple selection with markers for different traits provides advantages in selection efficiency. Additional markers may require very little additional effort in screening since the sampling and DNA preparation usually account for most of the work.

Conversion of markers such as RFLPs to PCR-based markers (STSs) may allow significant savings in the routine application of molecular markers in plant improvement (Talbert *et al.*, 1994).

The application of molecular markers to selection in plant breeding requires the availability of simple, inexpensive techniques that can provide rapid results to allow timely decisions on lines to be included in the next round of breeding and selection. Automation of sample handling, DNA extraction, DNA analysis, data capture and data analysis may be essential for application to large breeding populations. The need for automation may influence the choice of marker technique. Marker-assisted selection must always compete in cost and speed with the alternative of direct screening for the phenotype to be worthwhile. However, the ability to screen for additional markers at very little additional cost is a major consideration. If marker-assisted selection is justified for one or more traits, the sample handling, DNA extraction and data handling costs for the sample have been covered and screening for additional markers may cost very little. This may result in marker-assisted selection for some traits being an attractive option, despite the existence of simple alternatives.

3.4.1 INTRODUCTION

The quality of the product is a key target in many plant breeding programmes. The appearance, texture and flavour of whole foods such as fruit and vegetables may be the most important breeding objective. The processing quality of other species such as grain and fibre crops may be equally essential. Marker-assisted selection for quality provides

3.4 Marker-assisted breeding to improve product quality

an attractive approach to breeding for these characters because they are often influenced greatly by the environment, making highly replicated field trials necessary. Some quality attributes also require large samples for assessment, providing a further incentive for the application of marker-assisted selection. Most quality attributes are quantitative in nature, suggesting the involvement of many genes and complicating the application of markers. However, successful molecular analysis of many complex traits has indicated that they are often controlled by one or a few genes in any genotype. Individual genes may have a major influence on a quantitative character in a specific cross, while a much larger group of genes may contribute to the trait across the entire germplasm pool (Helentjaris, 1988).

The use of molecular markers may be of great value in the selection for desirable traits in long-lived species and those that take a long time to reach maturity and display the phenotype required. The breeding of forest trees and other tree crops may be a very slow process if mature tree characters such as timber quality or fruit quality need to be assessed for selection. For example, avocado (*Persea americana* Mill.) fruit quality has been associated with specific DNA fragments in genetic fingerprints (Mhameed *et al.*, 1995). Rapid evaluation of seedlings using molecular markers is an attractive option in these cases.

3.4.2 EXAMPLES

(a) QTLs for barley quality

The quality of double haploids from the cross between Steptoe and Morex were analysed in North American environments (Hayes *et al.*, 1993) (Figure 3.5). QTLs for α-amylase and β-amylase were found close to the location of the known function genes for these enzymes. Other quality-related QTLs were dispersed throughout the genome. A

Mapping genes in barley

The North American Barley Genome Mapping Project (NABGMP) is a collaboration among researchers in Canada and the US (Figure 3.5). The project has mapped Quantitative Trait Loci (QTL) for traits such as yield and malt extract. The project was originally based upon two crosses: Morex/Steptoe and Harrington/TR306. Mapping populations of 150 doubled-haploid lines were used for each cross. The traits were assessed in different field environments to allow analysis of linkages between QTLs and molecular markers.

Table 3.3 Quality traits for which markers have been identified in barley. (From Henry *et al.*, 1996.)

Trait	Population	Type of marker	Location of marker (Chromosome number)
Grain protein	S × M	RFLP	2,4,6,7
	H × T	RFLP	1,2,4,6,7
	D × M (winter)	RFLP	4,7
	D × M (spring)	RFLP	5,6,7
Kernel weight	S × M	RFLP	2,3,4,5,6,7
	S × M	RFLP	2,3,4,5,6,7
	H × T	RFLP	1,6,7
Plump grains	S × M	RFLP	2,3,4,5,7
	H × T	RFLP	2,4,5,6,7
Kernel length	I × D	RFLP	4,7
Kernel thickness	I × D	RFLP	7
Wort protein	S × M	RFLP	1,2,3,4,5,6,7
	H × T	RFLP	1,2,4,5,7
β-Glucan	S × M	RFLP	1,2,3,4,5,7
	H × T	RFLP	3,7
α-Amylase	S × M	RFLP	1,2,4,5,7
	H × T	RFLP	6,7
	D × M (winter)	RFLP	1,7
	D × M (spring)	RFLP	1,7
Diastatic power	S × M	RFLP	1,2,4,5,7
	H × T	RFLP	2,5,6,7
Malt extract	S × M	RFLP	1,2,4,5,7
	D × M (winter)	RFLP	2,3
	D × M (spring)	RFLP	7
Milling energy		RAPD	BSA

BSA, bulked segregant analysis.

QTL for diastatic power on chromosome 4 in Steptoe was not associated with any known function gene contributing to diastatic power suggesting that analysis of this locus may improve understanding of the genetic and molecular basis of these quality traits. Genes controlling milling energy of barley have also been mapped (Chalmers *et al.*, 1993). BSA was applied to analysis of a DH population from a cross between the cultivars Blenheim and a breeding line, E224/3. RAPD markers were found for a QTL for milling energy on the short arm of chromosome 5. Quality traits mapped in barley are listed in Table 3.3.

Chromosome 1

- Markers mapped in **Steptoe/Morex** are shown on the **left**
- Markers mapped in **Harrington/TR 306** on the **right**

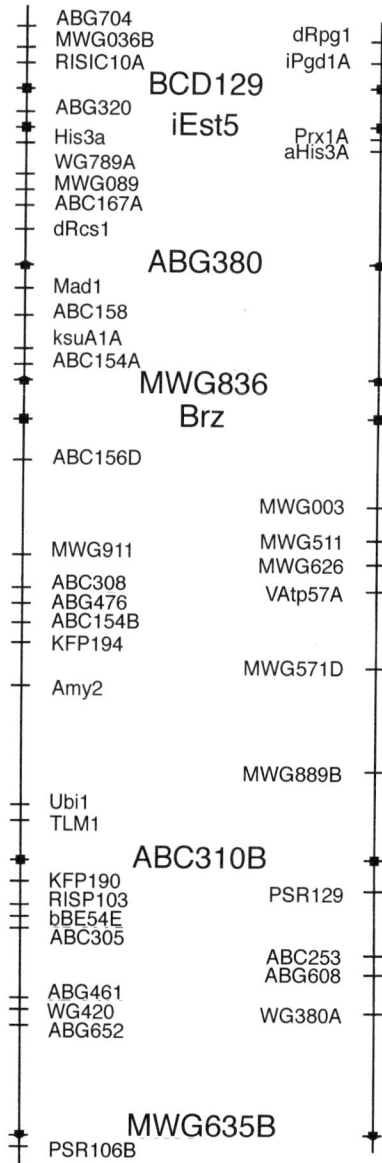

Figure 3.5 The North American Barley Genome Mapping Program (NABGMP).

Chromosome 2

- Markers mapped in **Steptoe/Morex** are shown on the **left**
- Markers mapped in **Harrington/TR 306** on the **right**

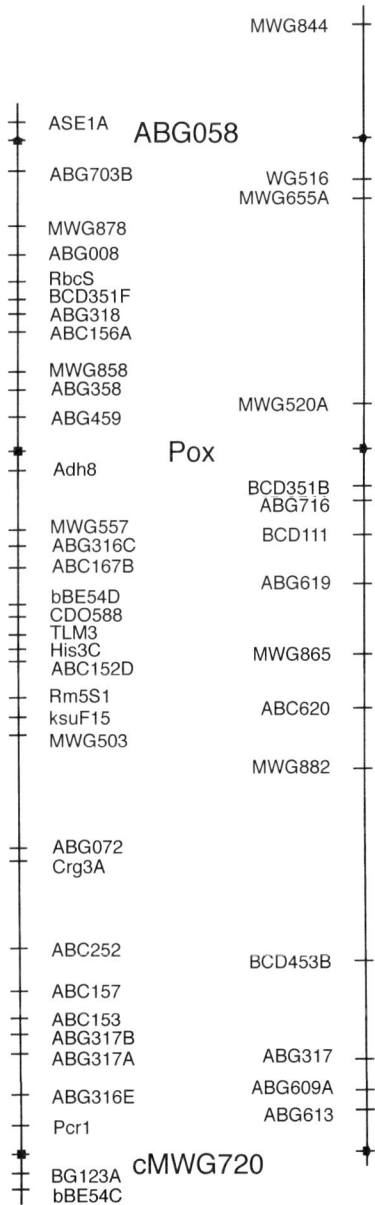

```
                                              MWG844  +


           +  ASE1A      ABG058
           •
              ABG703B                 WG516    +
                                    MWG655A    +
           +  MWG878
           +  ABG008
           +  RbcS
           +  BCD351F
           +  ABG318
           +  ABC156A
           +  MWG858
           +  ABG358
                                    MWG520A
           +  ABG459
           •                 Pox               •
           +  Adh8
                                    BCD351B    +
                                     ABG716
           +  MWG557         BCD111    +
           +  ABG316C
           +  ABC167B
                                     ABG619
           +  bBE54D
           +  CDO588
           +  TLM3
           +  His3C          MWG865    +
           +  ABC152D
           +  Rm5S1          ABC620    +
           +  ksuF15
           +  MWG503
                             MWG882    +


           +  ABG072
           +  Crg3A



           +  ABC252         BCD453B    +

           +  ABC157
           +  ABC153
           +  ABG317B
           +  ABG317A        ABG317     +
                             ABG609A    +
           +  ABG316E        ABG613
           +  Pcr1
           •          cMWG720            •
           +  BG123A
           +  bBE54C
```

Figure 3.5 Continued.

Chromosome 3

● Markers mapped in **Steptoe/Morex** are shown on the **left**
● Markers mapped in **Harrington/TR 306** on the **right**

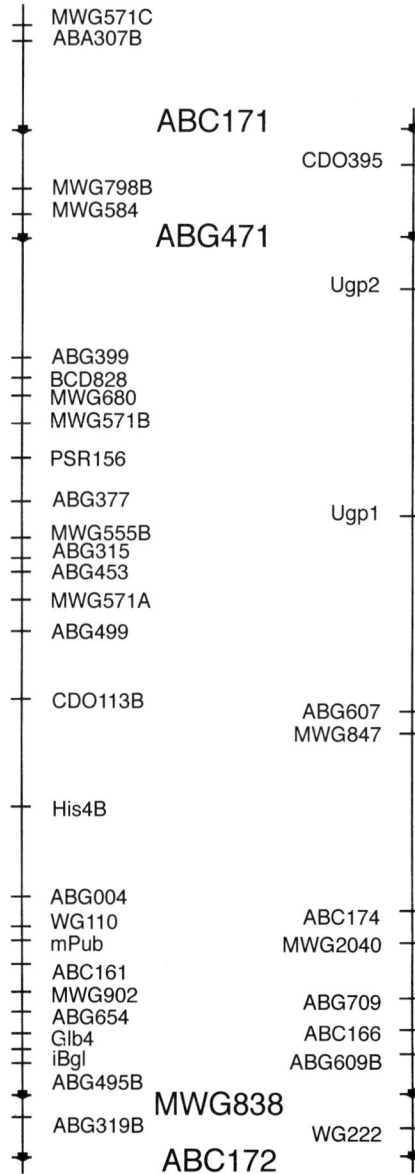

Figure 3.5 Continued

Chromosome 4

- Markers mapped in **Steptoe/Morex** are shown on the **left**
- Markers mapped in **Harrington/TR 306** on the **right**

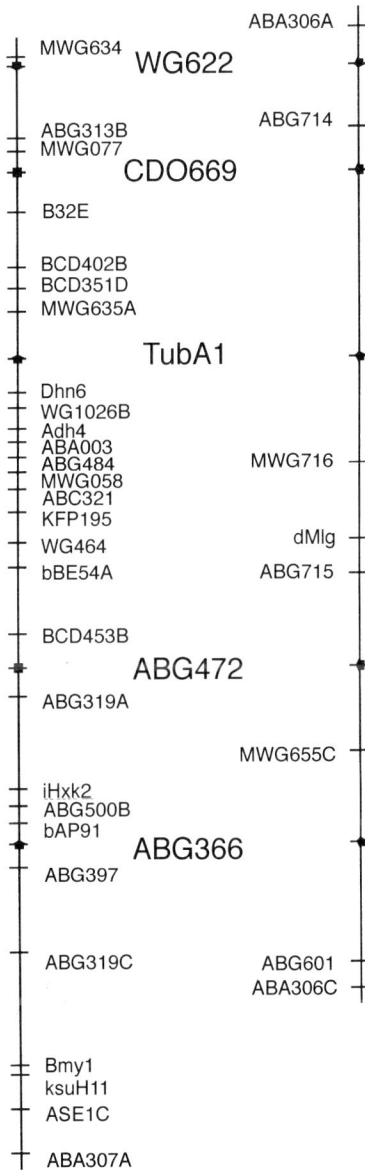

Figure 3.5 Continued.

Chromosome 5

- Markers mapped in **Steptoe/Morex** are shown on the **left**
- Markers mapped in **Harrington/TR 306** on the **right**

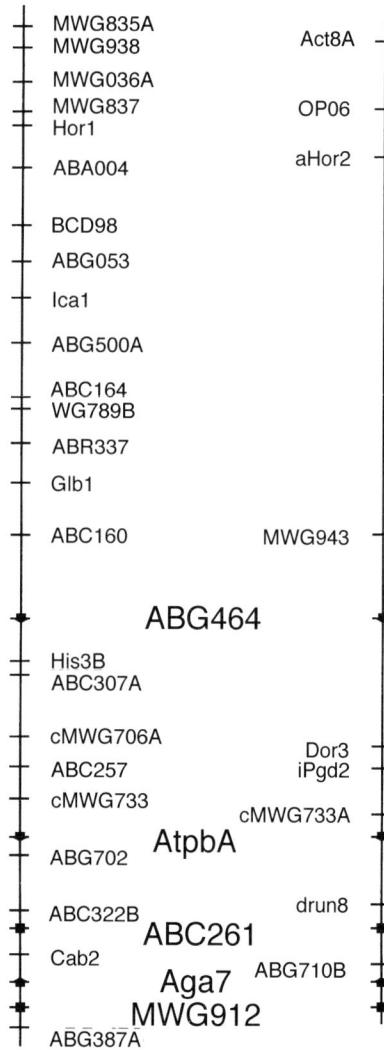

MWG835A
MWG938
MWG036A
MWG837
Hor1

ABA004

BCD98

ABG053

Ica1

ABG500A

ABC164
WG789B

ABR337

Glb1

ABC160

ABG464

His3B
ABC307A

cMWG706A
ABC257
cMWG733

AtpbA

ABG702

ABC322B
ABC261
Cab2
Aga7
MWG912
ABG387A

Act8A

OP06

aHor2

MWG943

Dor3
iPgd2

cMWG733A

drun8

ABG710B

Figure 3.5 Continued

Chromosome 6

- Markers mapped in **Steptoe/Morex** are shown on the **left**
- Markers mapped in **Harrington/TR 306** on the **right**

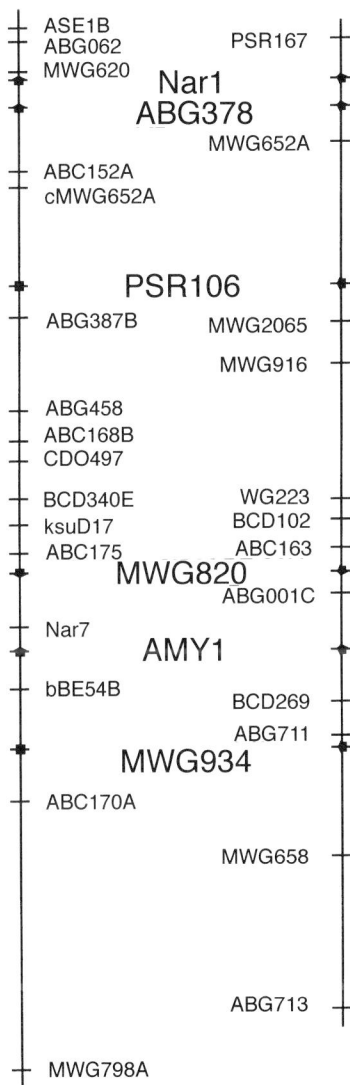

Left (Steptoe/Morex)	Center	Right (Harrington/TR 306)
ASE1B		PSR167
ABG062		
MWG620	Nar1	
	ABG378	
		MWG652A
ABC152A		
cMWG652A		
	PSR106	
ABG387B		MWG2065
		MWG916
ABG458		
ABC168B		
CDO497		
BCD340E		WG223
ksuD17		BCD102
ABC175		ABC163
	MWG820	ABG001C
Nar7		
	AMY1	
bBE54B		BCD269
		ABG711
	MWG934	
ABC170A		
		MWG658
		ABG713
MWG798A		

Figure 3.5 Continued.

Chromosome 7

● Markers mapped in **Steptoe/Morex** are shown on the **left**
● Markers mapped in **Harrington/TR 306** on the **right**

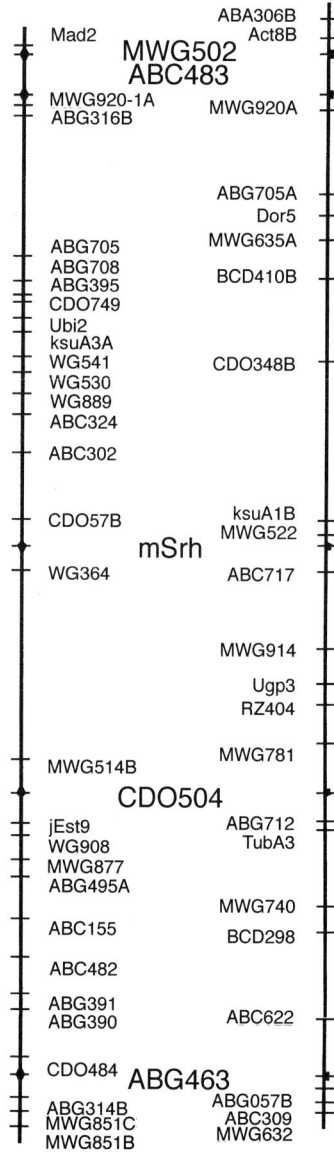

Figure 3.5 Continued

(b) Tomato fruit quality

Fruit quality traits have been introgressed into tomato (*Lycopersicon esculentum*) from the wild tomato (*Lycopersicon chmielewskii*) using a tightly linked RFLP marker for the sucrose accumulator gene (*surc*) (Chetelat *et al.*, 1995a). Selection for flanking markers was used to reduce the size of the chromosomal segment introduced (Chetelat *et al.*, 1995b). This approach may be used to manipulate the soluble solids content of tomatoes. The increased sucrose accumulation was associated with the production of larger numbers of smaller fruit but no change in overall yield. The proportion of ripe fruit at harvest was reduced, but the predicted paste yield was increased. A major QTL for fruit weight is present on chromosome 2 of red (*L. pimpinellifolium*) and green-fruited (*L. pennellii*) wild tomato (Alpert *et al.*, 1995). A similar QTL has also been reported at the same locus in *L. cheesmanii*. The identification of QTLs for the important fruit quality traits in tomato has several implications. The presence of this QTL in several species suggests that the QTL was present in ancestral species that predated the speciation events and indicates the potential value of comparative mapping in these plants. Marker-assisted selection for fruit quality may become an attractive option and map-based cloning of these genes may offer further options.

(c) QTLs for cucumber fruit quality

QTLs for the quality traits, length, diameter, seed-cavity size and colour were analysed in cucumber (*Cucumis sativus* L. var. *sativus*) (Kennard and Harvey, 1995). The numbers of QTLs their heritabilities and the proportion of the variation by the QTLS explained was analysed in backcrosses.

3.5.1 INTRODUCTION

Diseases are major constraints to crop production. Breeding of resistant plant varieties allows this limit on crop production to be overcome and is a crucial component of achieving production efficiency. Markers offer the possibility of selection in the absence of the pathogen, an option of considerable value in the breeding of plants with resistance to serious diseases that are not yet present in the region for which the plant is being developed.

3.5 Marker-assisted breeding for disease-resistant plants

3.5.2 EXAMPLES

(a) Barley stem rust

The *rpg4* gene confers resistance to stem rust (*Puccinia graminis* Pers. F. sp. *tritici* Erkis. & E. Henn) in barley. Bulked segregant analysis has been used to identify RAPD markers linked to this gene (Borovkova *et al.*, 1995). Markers from the established RFLP map of barley were used to locate the gene on chromosome 7. The markers allow the breeding of barley varieties with this important disease resistance gene.

(b) Rust of crucifers

Albugo canida causes white rust of the crucifers, radish (*Raphanus sativus*), brown mustard (*Brassica juneca*) and oilseed rape (*Brassica rapa*). A resistance locus has been mapped to linkage group 9 of rape by RFLP analysis of a F_1 double-haploid population segregating for a dominant single resistance allele (Ferreira *et al.*, 1995).

(c) Blast of rice

A sequence-tagged site for a genomic clone linked tightly to a blast resistance gene failed to amplify products specifically from genotypes with the resistance (Hittalmani *et al.*, 1995). However, the restriction enzyme, *Hae*III, generated a polymorphism when used to digest the product amplified by the STS and this polymorphism was linked to the resistance in 95% of the population tested.

(d) Head smut of sorghum

Head smut of sorghum is caused by the variable *Sporisorium reilianum* (Oh *et al.*, 1994). RFLP and RAPD markers linked to head smut resistance genes have been identified by screening of segregating F_2 populations and bulked segregant analysis.

(e) Root cyst nematode of potato

RFLP markers have been identified for genes conferring resistance to the root cyst nematode (*Globodwea rostochiensis*) in potato. The application of these markers has been simplified by the development of PCR markers for some of the resistance alleles (Niewohner *et al.*, 1995). The approach taken was to sequence the cDNA used as an RFLP probe for the target locus. The probe sequence allowed the use of PCR to explore sequence variations at the locus. PCR primers were eventually designed to detect specific alleles. This case illustrated the limitations of this approach by revealing that the alleles were not always linked to nematode resistance in other potato varieties.

(f) Cyst nematode of sugar beet

Screening of a YAC library of sugar beet (*Beta vulgaris*) carrying a gene for resistance to the cyst nematode (*Heterodera schachtii*) from a wild beet (*Beta procumbens*) allowed isolation of clones representing most of the translocation (Kleine *et al.*, 1995). Flanking RAPD markers for this resistance gene have also been established and converted to sequence tagged sites (Salentijn *et al.*, 1995).

(g) Mosaic virus of tomato

RAPD markers for genes conferring resistance to mosaic virus were identified by screening near isogenic lines of tomato (*Lycopersicon esculentum*). Markers linked to two different loci (*Tm-2* and *Tm-2a*) were shown to be closely linked to the netted virescent (*nv*) locus. These two genes were introduced from the wild relative (*Lycopersicon peruvianum*) and are located on the long arm of chromosome 9 (Ohmori *et al.*, 1995).

(h) Leaf mould of tomato

AFLP markers tightly linked to the *Cf-9* gene of tomato were identified by screening 42 000 AFLP loci (Thomas *et al.*, 1995). This gene confers resistance to the leaf mould pathogen, *Cladosporium fulvum*.

(i) Insect (gall midge) resistance in rice

RAPD markers tightly linked to gall midge (*Orseolia oryzae* Wood-Mason) resistance in rice have been identified for use in marker-assisted selection for midge resistance in the absence of the insect (Nair *et al.*, 1995).

(j) Rust resistance in sunflower

Sunflower (*Helianthus annuus* L.) is an important oilseed crop. In some areas, production is reduced by rust caused by *Puccinia helianthi* Schw. Sunflower genotypes grown commercially show good levels of DNA polymorphism (Lawson *et al.*, 1994) making marker-assisted selection an attractive option for sunflower breeders.

RAPD markers for sunflower rust resistance genes have been identified by bulk segregant analysis (Lawson *et al.*, 1996). A segregating F_2 population of 156 individuals was screened for resistance to a

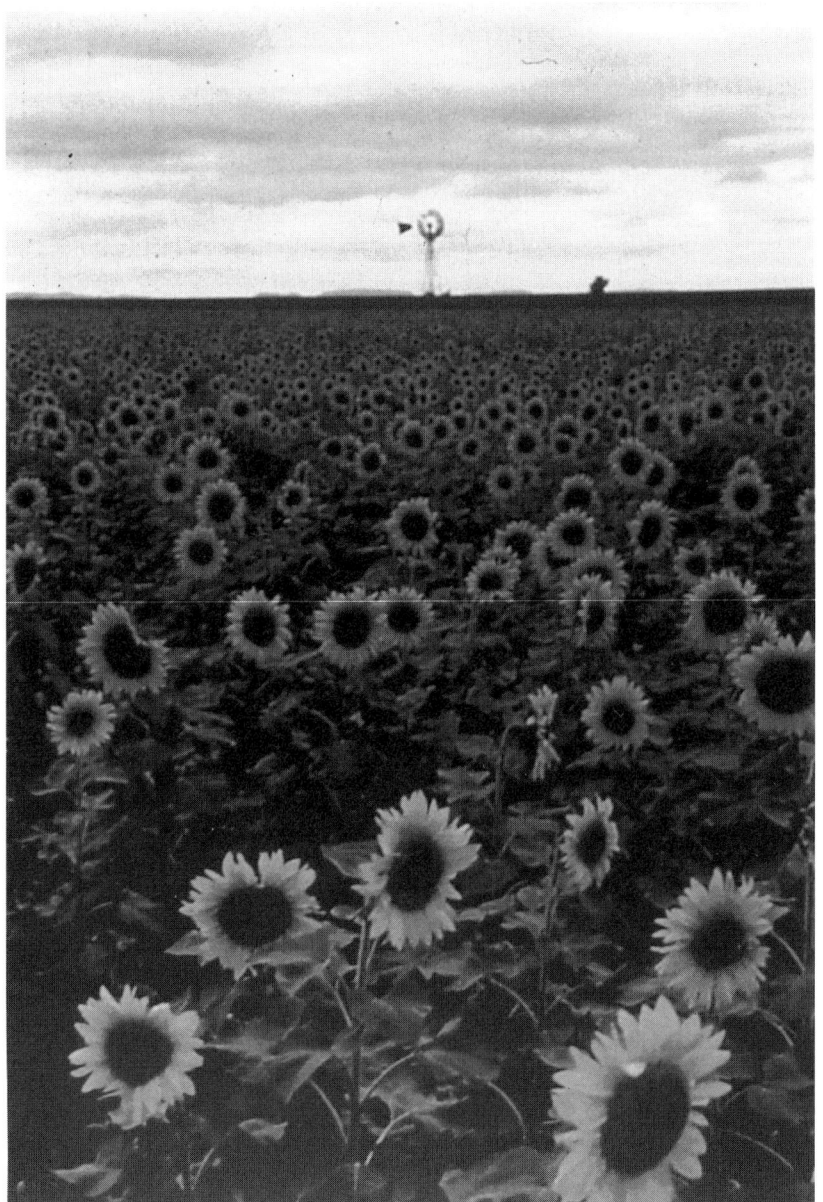

Figure 3.6 Sunflowers. Commercial sunflower genotypes show high levels of DNA polymorphism that can be used in marker-assisted selection.

specific race of this pathogen. Eight resistant and eight susceptible plants were used to produce DNA bulks. Screening of the two parental lines and the two bulks with 272 RAPD primers gave an average of five DNA markers per primer. Two primers resulted in a marker common to the resistant parent and resistant bulk, but absent in the susceptible parent and bulk. One marker $OT06_{950}$ was present in all eight resistant lines in the bulk and absent in all eight susceptible lines in the bulk. Screening of 76 lines from the mapping population indicated that two resistant individuals lacked the marker and one susceptible individual had the marker. Linkage analysis suggested that the marker and resistance gene were separated by 4.5 cm.

3.6.1 INTRODUCTION

A wide range of traits associated with plant performance may be targets for marker-assisted selection. Yield, and even components of yield, are usually quantitative traits. However, markers accounting for a significant proportion of yield can often be found. Other important plant traits in this category include plant height and maturity type (e.g. flowering time). A practical application of markers for these traits in plant improvement may be in identification of factors contributing to, for example, yield and the identification and conservation of germplasm for these traits.

> **3.6 Marker-assisted breeding for improved plant performance**

3.6.2 EXAMPLES

(a) Yield in maize

The QTLs for yield in a segregating F_2 population from an elite maize cross were analysed using RFLPs (Ajmone-Marsan *et al.*, 1995). The major QTL identified accounted for about one-quarter of the yield variation.

(b) Plant height in sorghum

Four QTLs accounting for 9–29% of variation in plant height in a segregating F_2 sorghum population were identified using RFLPs (Pereira and Lee, 1995). These four regions may correspond to those previously reported from maize.

(c) Plant height, time to heading and growth habit in barley

RAPD and RFLP markers have been applied to the location of genes controlling developmental traits in barley (Barua *et al.*, 1993). The *denso* dwarfing gene was mapped on the long arm of chromosome 3H. This gene was also found to contribute to delaying time to heading. Loci on chromosomes 5H and 6H contributed to heading time. Chromosome 7H also influenced height. The *denso* locus has contributed to resistance to lodging and increased harvest index.

KEY TERMS

Backcross (BC)
Bulked segregant analysis (BSA)
Co-dominant marker
Dominant marker
Doubled haploid (DH)
Heterosis
Linkage
Map
Quantitative trait loci (QTL)
Recombinant inbred line (RIL)
Synteny

EXAMPLES OF WORKED QUESTIONS

1. Define a doubled haploid.

   A doubled haploid is a homozygous line derived by the doubling of a haploid genome. This is usually achieved by anther culture or in some cases by the generation of a haploid in an incompatible pollination. Chromosome doubling in these lines may be spontaneous or induced.

2. How would you apply bulked segregant analysis to the identification of a molecular marker linked to a disease resistance gene?

   A cross between a resistant and susceptible line could be used to generate a population segregating for resistance. The population could be screened for resistance using appropriate isolates of the pathogen. Several (say eight) resistant and several (say eight) susceptible lines could be used to generate bulks. Equal

> amounts of DNA from these plants would provide two bulk
> DNA samples (resistant and susceptible). Screening of these bulks
> with molecular markers would reveal polymorphisms that might
> be linked to the resistance gene. Candidate polymorphisms can
> be verified by testing the plants making up the bulks individually
> and then establishing the linkage by applying the marker to the
> entire segregating population.
>
> 3. List some examples of practical applications of molecular
>   markers in plant breeding.
>
>   *Confirmation of the identity of breeding lines, establishment of
>   hybrid identity of lines, testing for purity of breeding lines,
>   evaluation of somaclonal variation, accelerated backcrossing and
>   selection for a desirable trait.*
>
> 4. List some of the factors to be considered in the choice of a
>   specific population for identification of a marker linked to
>   a desirable trait.
>
>   *Extent of variation in trait found in the population, size of
>   population available, level of polymorphism in the population,
>   availability of a map or markers known to be polymorphic in
>   the population.*

Questions

1. Define a recombinant inbred line.

2. How would you convert a RAPD marker identified by bulked
   segregant analysis to a SCAR for routine use in plant breeding?

3. Define the advantages provided by a molecular map of the species
   in the application of markers in plant breeding.

4. Describe how marker-assisted selection could be applied to accelerate the release of an apple with improved fruit quality.

References

Ajmone-Marsan, P., Monfredini, G., Ludwig, W.F., Melchinger, A.E., Franceschini, P., Pagnotto, G. and Motto, M. (1995) In an elite cross of maize a major quantitative trait locus controls one-fourth of the genetic variation for grain yield. *Theoretical and Applied Genetics*, 90, 415–24.

Alpert, K.B., Grandillo, SS. and Tanksley, S.D. (1995) *fw 2.2*: a major QTL controlling fruit weight is common to both red- and green-fruited tomato species. *Theoretical and Applied Genetics*, 91, 994–1000.

Barua, U.M., Chalmers, K.J., Thomas, W.T.B. and Hackett, C.A. (1993) Molecular mapping of genes determining height, time to heading, and growth habit in barley (*Hordeum vulgare*). *Genome*, **36**, 1080–7.

Berry, S.T., Leon, A.J., Hanfrey, C.C., Challis, P., Burkholz, A., Barnes, S.R., Rufener, G.K., Lee, M. and Caligari, P.D.S. (1995) Molecular marker analysis of *Helianthus annuus* L. 2. Construction of an RFLP linkage map for cultivated sunflower. *Theoretical and Applied Genetics*, **91**, 195–9.

Borovkova, I.G., Steffenson, D B.J., Jin, Y., Rasmussen, J.B., Kilian, A., Kleinhofs, A., Rossnagel, B.G. and Kao, K.N. (1995) Identification of molecular markers linked to the stem rust resistance gene *rpg*4 in barley. *Phytopathology*, **85**, 181–5.

Chalmers, K.J., Barua, U.M., Hackett, C.A., Thomas, W.T.B., Waugh, R. and Powell, W. (1993) Identification of RAPD markers linked to genetic factors controlling the milling energy requirement of barley. *Theoretical and Applied Genetics*, **87**, 314–20.

Chetelat, R.T., DeVerna, J.W. and Bennett, A.B. (1995a) Introgression into tomato (*Lycopersicon esculentum*) of the *L. chmielewskii* sucrose accumulator gene (*sucr*) controlling fruit sugar composition. *Theoretical and Applied Genetics*, **91**, 327–33.

Chetelat, R.T., DeVerna, J.W. and Bennett, A.B. (1995b) Effects of the *Lycopersicon chmielewskii* sucrose accumulator gene (*sucr*) on fruit yield and the quality parameters following introgression into tomato. *Theoretical and Applied Genetics*, **91**, 334–9.

Croy, R.D.D. (1993) *Plant Molecular Biology*, Labfax Bios Scientific Publishers, Oxford.

Dib, C., Faure, S., Fizames, C., Samson, D., Drouot, N., Vignal, A., Millasseau, P., Marc, L., Hazan, J., Seboun, E., Lathrop, M., Gyapay, G., Morissete, J. and Wessenbach, J. (1996) A comprehensive genetic map of the human genome based on 5,264 microsatellites. *Nature*, **380**, 152–4.

Ferreira, M.E., Williams, P.H. and Osborn, T.C. (1995) Mapping of a locus controlling resistance to *Albugo candida* in *Brassica napus* using molecular markers. *Phytopathology*, **85**, 218–20.

Huhn, M. (1995) Determining the linkage of disease-resistance genes to molecular markers: the LOD-SCORE method revisited with regard to necessary sample sizes. *Theoretical and Applied Genetics*, **90**, 841–6.

Hall, D.O. and Scurlock, J.M.O. (1993) Appendix C Biomass production and data, in *Photosynthesis and Production in a Changing Environment. A Field and Laboratory Manual* (eds D.O. Hall, J.M.O. Scurlock, H.R. Bolhar-Nordenkampf, R.C. Leegood and S.P. Long), Chapman & Hall, London, pp. 425–44.

Halward, T., Stalker, H.T. and Kochert, G. (1993) RFLP linkage map of diploid peanut (*Arachis sp.*) (2n = 2× = 20) genetic maps, in *Locus Maps of Complex Genomes; Book 6 Plants*, 6th edn (ed. S.J. O'Brien), Cold Spring Harbor Laboratory Press, Cold Spring Harbor, New York, pp. 86–7.

Hayes, P.M., Liu, B.H., Knapp, S.J., Chen, F., Jones, B., Blake, T., Franckowiak, J., Rasmusson, D., Sorrells, M., Ullrich, S.E., Wesenberg, D. and Kleinhofs, A. (1993) Quantitative trait locus effects and environmental interaction in a sample of North American barley germplasm. *Theoretical and Applied Genetics*, **87**, 392–401.

Helentjaris, T. (1988) Use of RFLP analysis to identify genes involved in complex traits of agronomic importance, in *Genetic Improvement of Agriculturally Important Crops Progress and Issues* (eds R.T. Fraley, N.M. Frey and J. Schell), Cold Spring Harbor Laboratory Press, Cold Spring Harbor, New York, pp. 27–30.

Henry, R.J., McLean, B.T. and Johnston, R.P. (1984) Determination of malting quality at different stages in a barley breeding programme. *Chemistry in Australia*, **51**, 247.

Henry, R.J., Weining, S. and Inkerman, P. (1996) Marker assisted selection for quality in barley and oat. International Barley Genetics Symposium (eds G. Scoles and B. Rossnagel), University Extension Press, University of Saskatchewan, Saskatoon, pp. 167–73.

Hittalmani, S., Foolad, M.R., Mew, T., Rodriguez, R.L. and Huang, N. (1995) Development of a PCR-based marker to identify rice blast resistance gene, *Pi-2(t)*, in a segregating population. *Theoretical and Applied Genetics*, **91**, 9–14.

Kennard, W.C. and Harvey, M.J. (1995) Quantitative trait analysis of fruit quality in cucumber: QTL detection, confirmation, and comparison with mating-design variation. *Theoretical and Applied Genetics*, **91**, 53–61.

Kleine, M., Cai, D., Elbl, C., Herrmann, R.G. and Jung, C. (1995) Physical mapping and cloning of a translocation in sugar beet (*Beta vulgaris* L.) carrying a gene for nematode (*Heterodera schachtii*) resistance from *B. procumbens*. *Theoretical and Applied Genetics*, **90**, 399–406.

Knapp, S.J., Holloway, J.L., Bridges, W.C. and Liu, B.-H. (1995) Mapping dominant markers using F_2 matings. *Theoretical and Applied Genetics*, **91**, 74–81.

Kurata, N., Nagamura, Y., Yamamoto, K. *et al.* (1994) A 300 kilobase interval genetic map of rice including 883 expressed sequences. *Nature Genetics*, **8**, 365–72

Larkin, P.J. and Scowcroft, W.R. (1981) Somaclonal variation – a novel source of variability from cell cultures for plant improvement. *Theoretical and Applied Genetics*, **60**, 197–214.

Lawson, W.R., Henry, R.J., Kochman, J.K. and Kong, G.A. (1994) Genetic diversity in sunflower (*Helianthus annuus* L.) as revealed by Random Amplified Polymorphic DNA analysis. *Australian Journal of Agricultural Research*, **45**, 1319–27.

Lawson, W.R., Goulter, K.C., Henry, R.J., Kong, G.A. and Kochman, J.K. (1996) RAPD markers for a sunflower rust resistance gene. *Australian Journal of Agricultural Research*, **47**, 395–401.

Mhameed, S., Hillel, J., Lahav, E., Sharon, D. and Lavi, U. (1995) Genetic association between DNA fingerprint fragments and loci controlling agriculturally important traits in avocado (*Persea americana* Mill.) *Euphitica*, **81**, 81–7.

Michellmore, R.W., Paran, I. and Kessli, R.V. (1991) Identification of markers linked to disease-resistance genes by bulked segregant analysis: a rapid method to detect markers in specific genomic regions by using segregating populations. *Proceedings of the National Academy of Sciences, USA*, **8**, 9828–32.

Moore, G., Gale, M.D., Kurata, N. and Flavell, R.B. (1993) Molecular analysis of small grain cereal genomes: current status and prospects. *Bio/Technology*, **11**, 584–9.

Nair, S., Bentur, J.S., Prasada Rao, U. and Mohan, M. (1995) DNA markers tightly linked to a gall midge resistant gene (Gm2) are potentially useful for marker-aided selection in rice breeding. *Theoretical and Applied Genetics*, **91**, 68–73.

Niewohner, J., Salamini, F. and Gebhardt, C. (1995) Development of PCR assays diagnostic for RFLP marker alleles closely linked to alleles *Gro1* and *H1*, conferring resistance to the root cyst nematode *Globodera rostochiensis* in potato. *Molecular Breeding*, **1**, 65–78.

O'Brien, S.J. (1993) *Genetic Maps Locus Maps of Complex Genomes; Book 6 Plants*, Cold Spring Harbor Laboratory Press, Cold Spring Harbor, New York.

Oh, J., Frederiksen, R.A. and Magill, C.W. (1994) Identification of molecular markers linked to head smut resistance gene (Shs) in sorghum by RFLP and RAPD analyses. *Phytopathology*, **84**, 830–3.

Ohmori, T., Murata, M. and Moyoyoshi, F. (1995) Identification of RAPD markers linked to the Tm-2 locus in tomato. *Theoretical and Applied Genetics*, **90**, 307–11.

Paran, I., Goldman, I., Tanksley, S.D. and Zamir, D. (1995) Recombinant inbred lines for genetic mapping in tomato. *Theoretical and Applied Genetics*, **90**, 542–8.

Paterson, A., Lin, Y.R., Li, Z., Schertz, K.F., Doebley, J.F., Pinson, S.R.M., Liu, S.-C., Stansel, J.W. and Irvine, J.E. (1995) Convergent domestication of cereal crops by independent mutations at corresponding genetic loci. *Science*, **269**, 1714–17.

Pereira, M.G. and Lee, M. (1995) Identification of genomic regions affecting plant height in sorghum and maize. *Theoretical and Applied Genetics*, **90**, 380–8.

Pimentel, D., Harvey, C., Resosudarmo, P., Sinclair, K., Kurz, D., McNair, M., Crist, S., Shpritz, L., Fitton, L., Saffouri, R. and Blair, R. (1995) Environmental and economic costs of soil erosion and conservation benefits. *Science*, **267**, 1117–23.

Poulsen, D.M.E., Ko, H.L., van der Meer, J.G., van der Putte, P.M. and Henry, R.J. (1996) Fast resolution of identification problems in seed production and plant breeding using molecular markers. *Australian Journal of Experimental Agriculture*, **36**, 571–6.

Ragot, M. and Hoisington, D. A. (1993) Molecular markers for plant breeding: comparisons of RFLP and RAPD genotyping costs. *Theoretical and Applied Genetics*, **86**, 975–84.

Salentijn, E.M.J., Aerens-De Reuver, M.J.B., Lange, W., De Block, Th.S.M., Stiekema, W.J. and Klein-Lankhorst, R.M. (1995) Isolation and characterization of RAPD-based markers linked to the beet cyst nematode resistance locus (*Hs1*[pat-1]) on chromosome 1 of *B. patellaris*. *Theoretical and Applied Genetics*, **90**, 885–91.

Schmidt, R., West, J., Love, K., Lenehan, Z., Lister, C., Thompson, H., Bouchez, D. and Dean, C. (1995) Physical map and organisation of *Arabidopsis thaliana* chromosome 4. *Science*, **270**, 480–3.

Schwarzacher, T. (1994) Mapping plants: progress and prospects. *Current Opinion in Genetics and Development*, **4**, 868–74.

Sorbral, B.W.S. and Honeycutt, R.J. (1993) High output genetic mapping of polyploids using PCR-generated markers. *Theoretical and Applied Genetics*, **86**, 105–12.

Talbert, L.E., Blake, N.K., Chee, P.W., Blake, T.K. and Magyar, G.M. (1994) Evaluation of 'sequenced-tagged-site' PCR products as molecular markers in wheat. *Theoretical and Applied Genetics*, **87**, 789–94.

Thomas, C.M., Vos, P., Zabeau, M., Jones, D.A., Norcott, K.A., Chadwick, B.P. and Jones, D.G. (1995) Identification of amplified restriction fragment polymorphism (AFLP) markers tightly linked to the tomato *Cf-9* gene for resistance to *Cladosporium fulvum*. *The Plant Journal*, **8**, 785–94.

Virk, P.S., Ford-LLoyd, B.V., Jackson, M.T., Pooni, H.S., Clemeno, T.P. and Newbury, H.J. (1996) Predicting quantitative variation within rice germplasm using molecular markers. *Heredity*, **76** (in press).

Williams, J.G.K., Reiter, R.S., Young, R.S. and Scolinik, P.A. (1993) Genetic mapping of mutations using phenotypic pools and mapped RAPD markers. *Nucleic Acids Research*, **21**, 2697–702.

Xu, Y., Clark, M.S. and Pehu, E. (1993) Use of RAPD markers to screen somatic hybrids between *Solanum tuberosum* and *S. brevidens*. *Plant Cell Reports*, **12**, 107–9.

Young, N.D. (1994) Constructing a plant genetic linkage map with DNA markers, in *DNA-Based Markers in Plants* (eds R.L. Phillips and I.K. Vasil), Kluwer Academic Publishers, Dordrecht, pp. 39–57.

Young, N.D. and Tanksley, S.D. (1989) Graphics-based whole genome selection using RFLPs, in *Development and Application of Molecular Markers to Problems in Plant Genetics* (eds T. Helentjaris and B. Burr), Cold Spring Harbor Laboratory Press, Cold Spring Harbor, New York, pp. 123–9.

Zhang, Q., Gao, Y.J., Saghai Maroof, M.A., Yang, S.H. and Li, J.X. (1995) Molecular divergence and hybrid performance in rice. *Molecular Breeding*, **1**, 133–42.

Plant transformation

4

Transgenic tobacco. Early experiments with easily transformed species such as tobacco have been followed by the development of many useful transgenic plants

After reading this chapter you should understand:

Techniques for Plant Transformation ● *Agrobacterium* Mediated Transformation ● Direct DNA Transfer ● Preparation of genes for Introduction into Plants ● Limitations of Transformation Techniques ● Study of Physiology and Biochemistry using Transformation ● Improvement of Product Quality using Transformation ● Nutritional Quality ● Functional Quality ● Control of Maturation in Fruit, Vegetables and Flowers ● Breadmaking Quality of Wheat ● Malting and Brewing Quality of Barley ● Storage Carbohydrate Composition of Plants ● Production of Pest and Disease Resistant Plants ● Insect Resistance ● Disease Resistance ● Bacteria and Fungi ● Viruses ● Nematodes ● Production of Herbicide Resistant Plants ● Herbicide Resistance as a Consequence of Transformation ● Herbicide Resistant Plants in Agriculture ● Production of Plants with Resistance to Abiotic Stress ● Production of Recombinant Proteins in Plants ● Use of Plants to Deliver Antigens ● Consumer Acceptance of Transgenic Plants ● Intellectual Property Issues ● Regulation of Transgenic Plants

> **Chapter outline**

The genetic engineering of plants provides an opportunity dramatically to alter the properties or performance of plants in order to improve their usefulness. Genetic engineering may be used to modify the expression of genes already present in the plant. New genes may be introduced from species with which the plant cannot be bred conventionally, and totally novel or synthetic genes may be added. This technology extends the possibilities of plant breeding to include totally new plant traits. The technology may also allow the achievement of aims of conventional plant breeding more efficiently. Single genes may be added to desirable genotypes without the alteration of any other features of the genotype that might often result from conventional backcrossing even with the assistance of molecular markers (Chapter 3). The techniques available for the genetic transformation of plants and their practical applications will be described in this chapter.

> **4.1 Introduction**

4.2.1 *AGROBACTERIUM*-MEDIATED TRANSFORMATION

The introduction of foreign DNA into plants may be achieved in several ways. The most effective approach in those plants amenable to the technique is probably *Agrobacterium*-mediated transformation (using *Agrobacterium tumefaciens* or *Agrobacterium rhizogenes*). *Agrobacterium*

> **4.2 Techniques for plant transformation**

tumefaciens may be viewed as a natural genetic engineer. As a soil pathogen this bacterium causes crown gall tumours in wounded plants. Gall-inducing strains contain a single copy of a plasmid (the transfer or Ti plasmid) that includes a segment (the transfer or T-DNA) that is stably incorporated into the plant genome. Vectors for plant transformation may be produced by replacing the T-DNA with the DNA to be introduced into the plant. The range of species that can be transformed using *Agrobacterium* is restricted but species previously considered outside the range of this technique have progressively been successfully transformed.

4.2.2 DIRECT DNA TRANSFER

Direct DNA transfer can be achieved by a range of physical methods. Microparticle bombardment (Birch and Bower, 1994), electroporation, polyethylene glycol (PEG) treatment, abrasion with fibres, micro-injection and laser-mediated (Guo *et al.*, 1995) approaches have proven successful in different cases. A wide range of these methods, especially electroporation and PEG-mediated transformation, have been success-fully used to introduce DNA into plant protoplasts. However, proto-plast regeneration to whole plants is difficult especially for some species and is often very genotype-dependent. The most widely used technique for cell and organ cultures is microparticle bombardment. The DNA is coated on tungsten or gold microparticles (approximately 1 μm) and projected into the cells to be transformed using a 'gun' powered with gunpowder or an inert gas (usually helium). This technique allows the cell wall to be penetrated and provides a more generally successful approach. A wide range of species has been successfully transformed using this technique (Sagi *et al.*, 1995). Fibres coated with DNA repre-sent a variation on this approach, vortex mixing of the suspended plant cells and fibres results in transformation associated with penetration of the cell membrane by the fibres. Lasers provide another mechanism for allowing DNA uptake through the cell membrane. Introduction of DNA into pollen provides a route for DNA transmission into seeds. Electro-poration may be used to transfer DNA into pollen and pollination with this pollen may generate transgenic plants (Smith *et al.*, 1994). DNA may be targeted to specific tissues or cell types by microtargeting (Gisel *et al.*, 1996). This approach may allow the specific transformation of cells that are totipotent (e.g. meristem cells).

4.2.3 PREPARING GENES FOR INTRODUCTION INTO PLANTS

Successful expression of foreign genes in plants requires the preparation of a suitable gene construct for introduction into the plant. In addition to the sequences encoding the gene product, appropriate promoter and termination sequences must be added at the 5' and 3' ends. The promoter requires a site for initiation of transcription with regulatory sequences to ensure the desired tissue and developmental patterns of expression. The correct 5' sequence alone will not necessarily ensure effective expression. For example, D'Ovidio *et al.* (1996) found no consistent differences in the 5' sequences of active and inactive glutenin subunits in wheat. Other features of the gene may prevent expression. Termination sequences are necessary to ensure effective termination of transcription. The coding region requires a translation initiation and termination codon and care is required to maintain the codons in frame. As a further refinement, the codon usage may need to be adjusted to maximize plant expression. Introns may also be used to enhance expression. The presence of introns may have different effects in different plants. For example, Li *et al.* (1995) reported that the insertion of an intron between the CaMV 35S promoter and GUS increased expression 15-fold in rice but not at all in tobacco protoplasts.

4.2.4 LIMITATIONS OF TRANSFORMATION TECHNIQUES

A growing list of plant species have been successfully transformed. However, many technical problems associated with plant transformation remain to be resolved.

Many of the direct DNA transfer methods are highly genotype-dependent because of the need for tissue culture. Improvement of genotype-independent cell culture and plant regeneration techniques (Figure 4.1) are required to allow more general application of plant transformation techniques.

Protocols are now available for the transformation of specific species or genotypes (Figure 4.2).

Gene expression may not be stable in some systems because of gene silencing, especially of homologous genes (Matzke and Matzke, 1995). Techniques for reliable gene targeting in plants may assist in overcoming this problem. Ideally, techniques are needed for the routine insertion of genes at precise loci and with control of the flanking sequences. Genes inserted into chloroplasts are mainly incorporated by homologous recombination but nuclear genes may also be targeted by using homologous recombination (Offringa and Hooykaas, 1995). Loss of expression may be due to methylation (up to 30% of cytosines in plants may be methylated as 5-methylcytosine) and differential methylation may explain differences in the expression of transgenes (Lambe *et al.*, 1995).

Gene replacement by homologous recombination would allow the elimination of undesirable genes by direct replacement or the

Figure 4.1 Regenerating transgenic cereal plants in a glasshouse. The successful regeneration of transgenic plants may be a limiting step in genetic engineering.

modification of gene function. The success of these methods is likely to be influenced by factors such as the length of homology between the vector and the target (Morton and Hooykaas, 1995).

The availability of suitable promoters (Table 4.1) may restrict some applications. Very precise control of the tissue and developmental stage at which the gene is expressed may be required to achieve the desired plant performance.

Direct DNA transfer methods do not allow control of the number of gene copies inserted requiring careful selection from a population of transformants. *Agrobacterium*-based transformation results in a high frequency of single copy transformants.

The production of transgenic plants usually involves co-transformation with a selectable marker gene (Table 4.2). This gene is in most cases an antibiotic resistance gene or a herbicide resistance gene. These genes may not be desirable in the transgenic plant. Non-selectable marker genes or reporter genes (Table 4.3) may be used to optimize transformation (McKinnon *et al.*, 1996) but these genes do not allow selection of transformed cells on a medium containing a selection agent. Better options for selectable marker genes could allow the production of transgenic plants more acceptable to consumers.

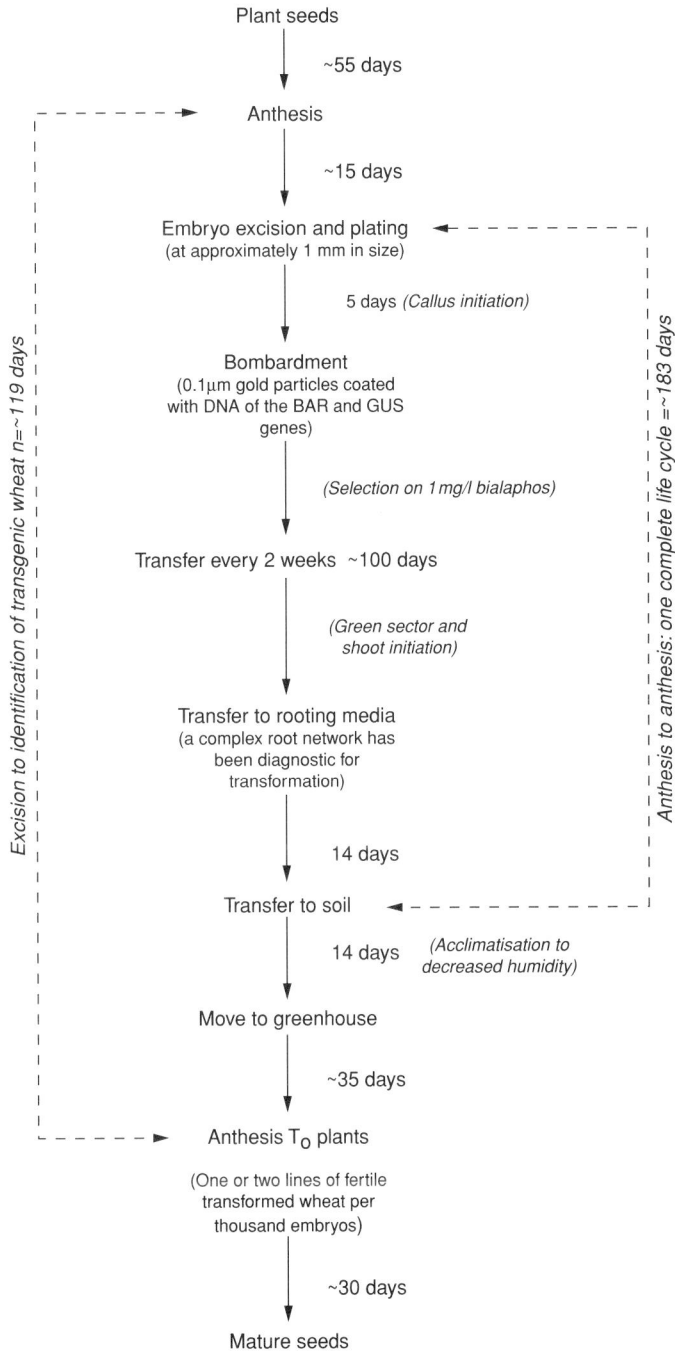

Figure 4.2 Wheat transformation protocol used by Anderson *et al.* (1994).

Table 4.1 Summary of constitutive promoters that have been used in transgenic cereals. (From McElroy and Brettall, 1994.)

Promoter	Source	Relative activity in cereal cells	Use in transgenic cereals
35S	Cauliflower mosaic virus 35S RNA transcript	Low	Rice Maize Fescue
35S-*Adh1* intron 1	Cauliflower mosaic virus 35S promoter and first intron of maize alcohol dehydrogenase 1 gene	Low	Maize Oats Wheat
Emu	Modified maize alcohol dehydrogenase 1 promoter and first intron	Moderate	Sugar cane Rice
Act1-Act1 intron 1	Rice actin 1 gene	Moderate	Rice
Ubil-Ubil intron 1	Maize ubiquitin 1 gene	High	Wheat Barley Rice

4.3 Using transformation to study plant physiology and biochemistry

The manipulation of plant performance may be achieved by the alteration of specific metabolic pathways using transformation. However, this requires a detailed knowledge of the physiology and biochemistry that determine plant performance. Transformation may be used as a tool to analyse these processes. Enhancing or restricting the metabolic flux through specific steps can be achieved by overexpressing a gene encoding the enzyme responsible, or by using antisense strategies to block production of the enzyme. The size of the pools of specific metabolites can be manipulated to improve plant performance or product quality. However, pathways of carbon or nitrogen metabolism need to be evaluated in detail before any attempts to apply these approaches in order to ensure the desired result.

Attempts to alter plant metabolism using native plant genes may meet limited success as expression of higher levels of key enzymes may not alter metabolic flux because the pathway is regulated. Several mechanisms may exist in the plant which allow compensating production of other enzymes, or direct metabolic regulation of the activity of the enzyme being overproduced. Expression of proteins from heterologous systems may avoid this limitation if, for example, the enzyme is free from the allosteric regulation that controls the plant enzyme.

Table 4.2 Selectable marker genes for use in plant transformation. (From Yoder and Goldsborough, 1994.)

Marker gene	Gene product	Source of gene	Selection agent
npt1I	Neomycin phosphotransferase	Tn5	kanamycin, G418, paromomycin, neomycin
Ble	Bleomycin resistance	Tn5 and *Streptoalloteichus hindustanus*	Bleomycin, phleomycin
dhfr	Dihydrofolate reductase	plasmid R67	Methotrexate
cat	Chloramphenicol acetyl transferase	phage p1 Cm	Chloramphenicol
aphIV	Hygromycin phospho-transferase	*E. coli*	Hygromycin B
SPT	Streptomycin phospho-transferase	Tn5	Streptomycin
aacC3, *aacC4*	Gentamycin-3-*N*-acetyl-transferase	*Serratia marcescens*; *Klensiella pneumoniae*	Gentamycin
bar	Phosphinothricin acetyl transferase	*Streptomyces hygroscopicus*	Phosphinothricin, bialophos
EDSP	5-Enolpyruvylshikimate-3-phosphate synthase	*Petunia hybrida*	Glyphosate
bxn	Bromoxynil specific nitrilase	*Klebsiella ozaenae*	Bromoxynil
psbA	Q_B protein	*Amaranthus hybridus*	Atrazine
tfdA	2,4-D monooxygenase	*Alcaligenes eutophus*	2,4-Dichloro-phenoxyacetic acid
DHPS	Dihydrodipicolinate synthase	*E, coli*	S-Aminoethyl-L-cysteine
AK	Aspartate kinase	*E. coli*	High concentrations of lysine and threonine
sul	Dihydrodipicolinate synthase	plasmid R46	Sulphonamide
csrl-1	Acetolactate synthase	*Arabidopsis thaliana*	Sulphonylurea herbicides
tdc	Tryptophan decarboxylase	*Catharanthus roseus*	4-Methyl tryptophan

Table 4.3 Reporter genes that have been used in plant transformation. (Modified from McElroy and Bretell, 1994.)

Properties	β-Glucuronidase	Luciferase	Anthyocyanin regulators	Green fluorescent
Source	*E. coli*	Firefly	Maize	Jellyfish
Background activity in plants	Low (some cases due to bacterial contaminants)	Low	Low–moderate (depending upon species/ tissues)	Low
Nature of assay	Destructive	Non-destructive	Non-destructive	Non-destructive
Stability of assay	High	Low	Low	High
Sensitivity of assay	Good	Moderate	Low	High
Simplicity of assay	Good	Poor	Good	Good
Quantitive nature of assay	Good	Good	Moderate	Moderate
Adverse effects on transgenic plant metabolism	Low	Low	High	High (at high levels of expression)
Relative cost of assay systems	Moderate	High (requires expensive detection equipment)	Low	Low

---

4.4 Improving product quality by transformation

4.4.1 INTRODUCTION

Plant transformation may be viewed as offering only two practical outcomes: (i) improved plant productivity; or (ii) improved plant product quality. The relative value of the two outcomes will depend upon the competitiveness of the market and the constraints on production efficiency. Improvements in quality will be important in commodities in which there is strong market competition. Transformation may be used to introduce new or novel characteristics that create a new market or displace conventional products. The improvements may relate to the nutritional value of the plant or the functional properties in processing or consumption.

The regulation and study of plant metabolism

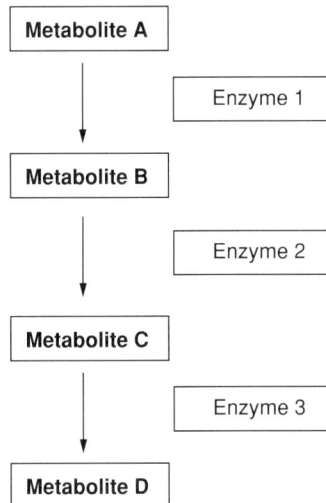

Metabolite A

Enzyme 1

Metabolite B

Enzyme 2

Metabolite C

Enzyme 3

Metabolite D

The level of metabolite D may be controlled by the overexpression or inhibition (e.g. antisense expression) of any of the three enzymes. However, enzyme 3 might be the best target. Control of the total flux though the pathway is probably best achieved by regulating enzyme 1. The level of metabolite B could be influenced by either enzyme 1 or enzyme 2.

4.4.2 NUTRITIONAL QUALITY

The nutritional value of foods may be enhanced by using genetic engineering to alter the composition of the edible part of the plant. Changes in the storage life of fruit and vegetables, for example, may improve their nutritional value at the time of consumption. The carbohydrate, protein, fat, fibre and vitamin content all have the potential to be altered. The amino acid composition of proteins may be adjusted so that the proteins have a higher nutritional value (Shewry *et al.*, 1994). The low lysine and threonine of cereals such as wheat and barley limits their nutritional value for humans and monogastric animals (e.g. pigs and poultry). The expression of proteins with higher levels of these amino acids would improve the nutritional value of these grains. Alterations of grain protein composition may be complicated by the impact of such changes on other aspects of grain quality. For example, the opaque 2 mutation in maize is associated with a significant increase in the lysine content but also results in an unacceptable soft texture. The most abundant proteins in cereals often have a major influence on quality and are those with the greatest imbalance in amino acids. Similar approaches may be useful for improving the nutritional value of other staple crops such as potato (Destefano-Beltran *et al.*, 1991). The antinutritional factors such as protease inhibitors and haemaglutinins in legumes and other plants may be reduced by transgenic approaches (Lumen, 1990). Problems associated with flatulence in some foods may be addressed by manipulating the dietary fibre and oligosaccharide content.

4.4.3 FUNCTIONAL QUALITY

The quality of food products may be improved by transformation of the plant from which the food is derived. The production of fruit and vegetables with improved flavour and texture by manipulation of maturation is a good example of this application and will be described here. The performance of raw plant products during processing may be improved by genetic engineering. Several examples will be outlined here: improvements in the quality of cereals (Henry and Ronalds, 1994) such as the breadmaking quality of wheat and the brewing quality of barley, and manipulation of the storage carbohydrate composition of plants to enhance processing.

(a) Control of maturation in fruit, vegetables and flowers

The maturation of fruit and vegetables (Figure 4.3) may involve changes in the sugars and acids contributing to the taste of the food, the softening of the edible parts of the plant, and other changes in factors contributing to flavour. The rate of these processes both before and after harvest may determine the quality of the food as perceived by the consumer.

The first genetically engineered whole food to be produced commercially was the FLAVR SAVR tomato (Kramer and Redenbaugh, 1994). This product is a tomato with an antisense gene blocking the production of polygalacturonase during fruit ripening. Polygalacturonase is a key enzyme involved in the degradation of pectic components of the cell walls in plants. Fruit softening during ripening of tomatoes has been inhibited by the expression of an antisense RNA to tomato polygalacturonase.

Figure 4.3 Fruit quality and shelf-life may be enhanced by transformation. The rate of maturation and the texture of the fruit may be modified.

A more general strategy for the control of ripening is to control the production of the ripening hormone, ethylene. This approach has potential to improve the shelf-life of a wide range of fruits. Ethylene is produced from S-adenosylmethionine by conversion to 1-amino-cyclopropane-1-carboxylic acid (ACC) under the control of ACC synthase, followed by the generation of ethylene by an ACC oxidase or ethylene-forming enzyme (EFE) (Figure 4.4). Antisense constructs directed against either of these enzymes, or removal of ACC with an ACC deaminase, may be used to block ripening (Romano and Klee, 1993). This approach may be used to produce fruit, vegetables and flowers with greatly enhanced shelf-life. Fruit may then be ripened as required by exposure to an artificial source of ethylene.

Ethylene biosynthesis

Methionine   (Met)

                                                    Ado Met Synthase

S-adenosylmethionine   (Ado Met)

                                                    ACC synthase

1-amino-cyclopropane-1-carboxylic acid   (ACC)

                                                    ACC Oxidase

                        Ethylene

                                                    $(CH_2=CH_2)$

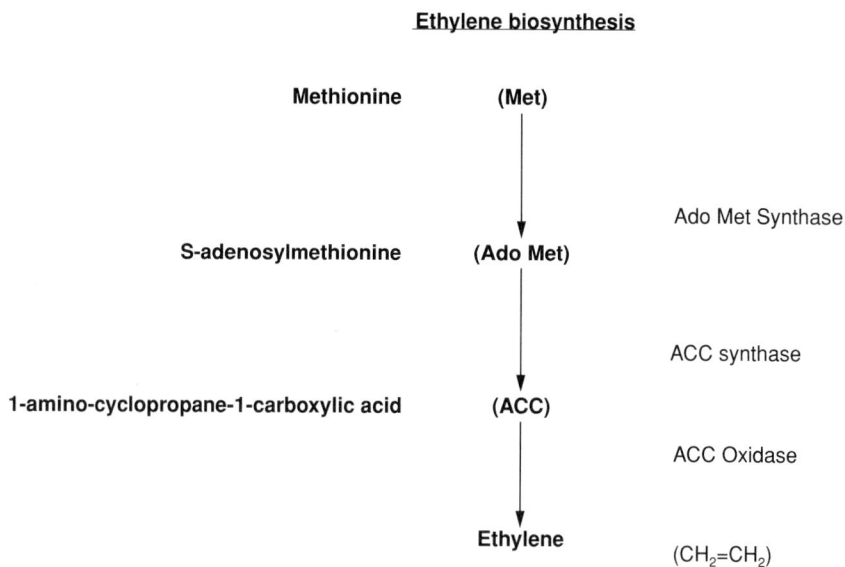

Figure 4.4 Biosynthesis of ethylene. Inhibition of this pathway by transformation allows control of ripening and senescence in plants.

(b) Breadmaking quality of wheat

The breadmaking quality of wheat (Figure 4.5) is largely attributed to the properties of the storage proteins of the grain (Shewry *et al.*, 1995). Genetic engineering of wheat quality may be achieved by manipulating the quantity and types of proteins in the grain. The high-molecular weight glutenin subunits of wheat have been well characterized and are the target of attempts to improve breadmaking quality. These proteins are responsible for the unique properties of a dough made from wheat flour. The gluten proteins allow stretching of the dough mixture to trap carbon dioxide bubbles (produced by the yeast fermenting carbohydrates in the dough) that can be stabilized by baking, to produce the crumb structure of products such as bread. The exact features of the proteins that contribute to these properties may be better defined by the study of different combinations of these proteins and mutated proteins in transgenic wheats. The formation of dough with elastic properties is dependent upon the formation of polymers of the high-molecular weight gluten subunits, stabilized by interchain disulphide bonds. Testing of natural and synthetic subunits of different sizes and with different numbers of cysteines in transgenic wheat plants will allow proteins with optimal properties to be designed for use in breadmaking wheats. Environmental problems associated

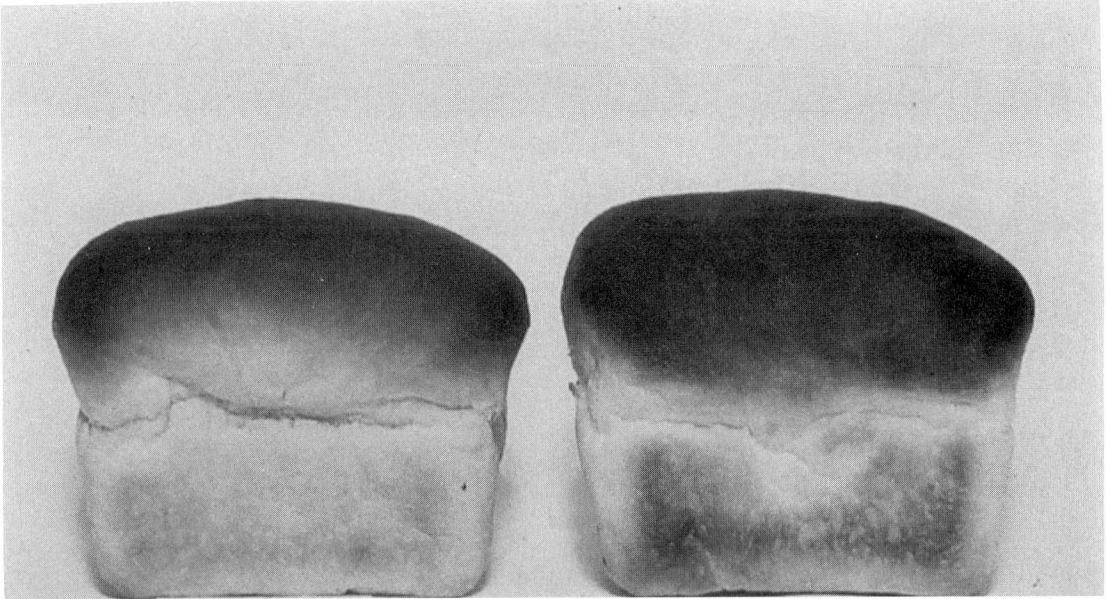

Figure 4.5 Breadmaking quality of wheat may be manipulated by transformation. Specific proteins contribute to breadmaking quality and may be expressed at higher levels or in different combinations. The size (volume), crumb texture and other qualities of loaves of bread are routinely evaluated by small-scale or test baking. Even small loaves contain many grams of flour, usually requiring the milling of flour from more than a single transgenic plant.

with the use of excessive amounts of nitrogenous fertilisers provides an incentive to develop genotypes that will allow breadmaking with wheats at the lowest possible protein content. This requires that as much of the protein as possible contributes to desirable dough properties.

The starch properties of wheat may also be altered to advantage, especially for the production of products such as noodles and for starch production. Changes in the proportion of amylose, the size distribution of the granules, and the fine structure of branching in the amylopectin may all be evaluated for their value in processing quality using transgenic wheats.

(c) Malting and brewing quality of barley

Beer production (Figure 4.6) involves the germination of barley under controlled conditions (malting), the production of a solution of sugars

Figure 4.6 Beer production is a large-scale industrial process converting raw materials of plant origin (barley) into a beverage (beer). Manipulation of the composition of barley has the potential to improve the ease of processing. The large scale of the process makes even small gains worthwhile.

by extraction of the ground malt in hot water (mashing), and the fermentation of the solution (wort) (Figure 4.7). The composition of the barley grain is an important factor determining the efficiency of the malting and brewing process and the quality of the finished beer. Many components of the grain have been proposed as targets for the genetic engineering of barley to improve its suitability for beer produc-

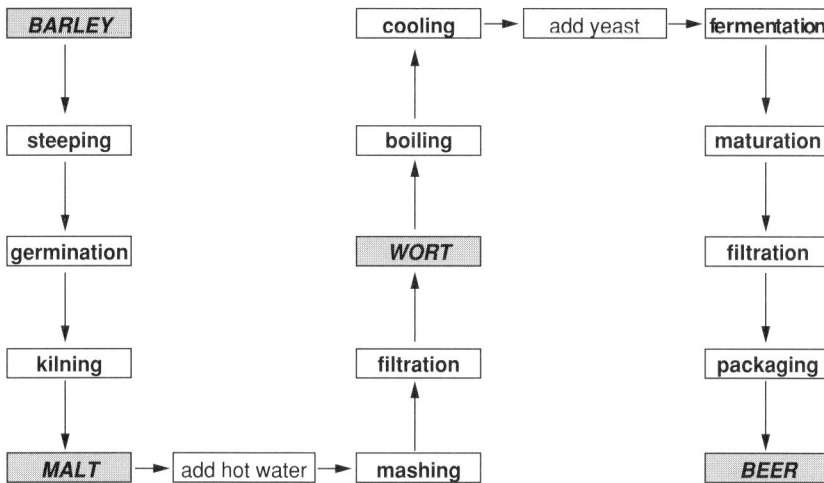

Figure 4.7 Steps in beer production.

tion (McElroy and Jacobsen, 1995; MacGregor, 1996) (Table 4.4). Candidates for manipulation include the major substrates for yeast fermentation and minor components influencing beer flavour. The manipulation of the activities or properties of enzymes from the plant has also been suggested. Increasing the stability of some of these enzymes at high temperatures may increase their effectiveness at the temperatures used in mashing (the first step in brewing). Vickers *et al.* (1996) evaluated the potential for the heat-stable α-amylase from *Bacillus licheniformis* to contribute to improved mashing. Expression of this bacterial enzyme in malt at about 0.5% of total protein would approximately double the activity of α-amylase in the mash, resulting in more efficient conversion of starch to sugars. The properties and relative activities of other starch-degrading enzymes in malt may also be worth manipulation. Other enzymes (e.g. β-amylase, α-glucosidase and limit dextrinase) are much less stable at brewing temperatures, suggesting potential for even more dramatic effects on the brewing processing following alteration of the properties of these enzymes.

Increased stability of β-glucanases at brewing temperatures may reduce the levels of β-glucans in the wort, improving filtration and potentially also beer stability. This may be achieved by modifying the properties of the barley proteins by gene replacement or introducing genes from other sources such as *Trichoderma reesi* (MacGregor, 1996).

Table 4.4 Targets for genetic improvement of barley as a raw material for brewing*

Carbohydrates	Starch	• higher malt extract • increased fermentability
Cell walls	β-Glucan	• better filtration • improved shelf-life
Proteins	Hordeins	• improved shelf-life • increased foam stability
	Enzymes[†]	• more efficient mashing • better filtration • increased fermentability
Other	Lipoxygenases Proanthocyanidins	• reduced off-flavours • reduced hazes

*Transformation may provide options for addressing each of the targets listed.
[†]Increased enzyme levels or improved enzyme stability at brewing temperatures may be achieved.

The flavour of beer could be manipulated by genetic engineering of the barley to reduce the risk of off-flavours. One such approach involves reducing the activity of lipoxygenases (Figure 4.8) to limit the level of the undesirable *trans*-2-nonenal formed by oxidation of barley lipids.

Inhibition of the formation of dimethyl sulphide (DMS) (Figure 4.9) is an attractive option for barleys to be used in lager production (Baxter, 1995).

(d) Storage carbohydrate composition of plants

Enhancement of starch synthesis may be achieved by expression of increased levels of ADP glucose pyrophosphorylase (Stark *et al.*, 1992). This may contribute to enhanced yields of starchy foods. The properties of plant starches may also be altered by plant transformation. The proportion of amylose and amylopectin in starch and the extent of branching may be regulated by transformation. This allows the tailoring of starches to meet specific requirements for specific food or industrial products.

Transgenic plants with elevated levels of fructans have been produced using a levansucrase from bacteria (Ebskamp *et al.*, 1994). Fructans may be hydrolysed to generate the intensely sweet sugar, fructose. Maize starch is hydrolysed to release glucose and this is converted to fructose enzymatically on an industrial scale to produce food sweeteners. Fructose production from fructan-accumulating transgenic plants may be an attractive alternative.

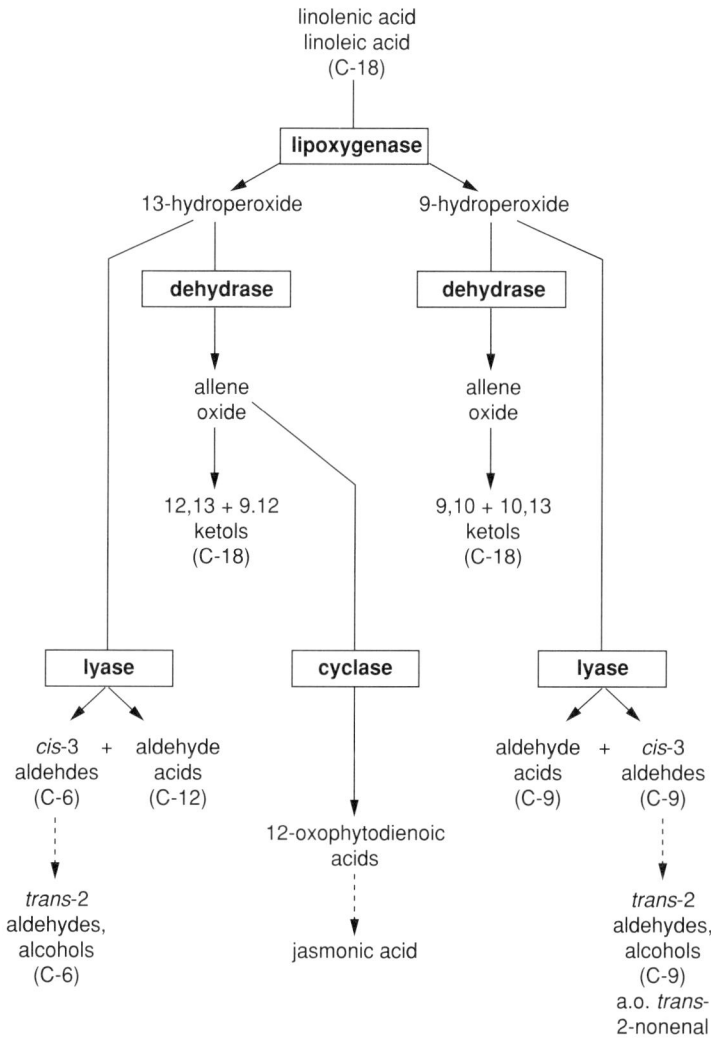

Figure 4.8 The lipoxygenase pathway in plants. Inhibition of these enzymes may improve the flavour of beer.

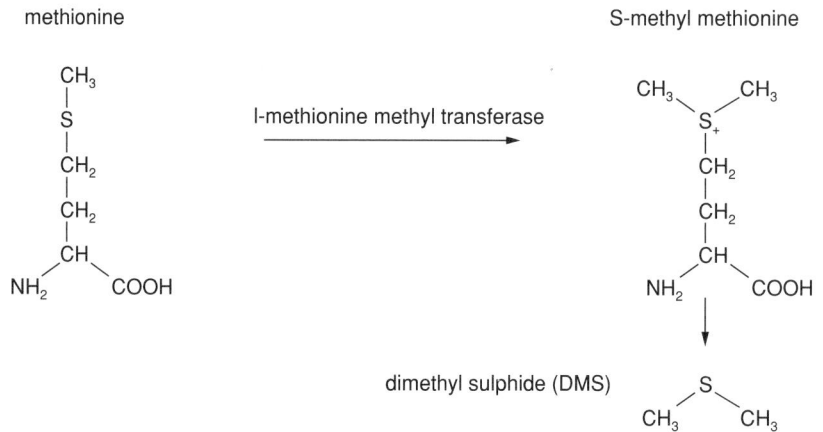

Figure 4.9 The formation of dimethyl sulphide (DMS) in beer production. DMS levels influence beer quality.

Fructans produced from sucrose-using enzymes are also used directly as low-calorie sweeteners in foods. Transgenic plants may be a useful source of fructans for this application.

The sucrose content of plants may also be manipulated to advantage in fruits to enhance quality and sugar crops such as sugarcane and sugar beet to improve yields of sugar.

(e) Other quality characteristics

Many other quality characteristics may be manipulated by transformation. For example, regulation of polyphenol oxidases may reduce discoloration of plants (e.g. potato) in storage and processing. Control of lipoxygenases may reduce the beany flavour of many pulses. The colour of flowers and of fruit and vegetables can be manipulated by control of the expression of genes involved in the regulation of anthocyanin synthesis (Holton and Cornish, 1995). Ornamental flower colour ranges can be extended using this approach. The appearance of fruit such as apples and grapes may also be modified.

The composition of plants could also be altered to improve their suitability as feedstocks for the production of industrial products such as biodegradable plastics (Nawrath *et al.*, 1995).

A major strategy for the improvement of plant production efficiency is to reduce losses due to pests and diseases (Figure 4.10) (Ward *et al.*, 1994). These problems persist post-harvest, contributing to further losses in production quantity and quality.

4.5.1 INSECT RESISTANCE

The environmental and health implications of the use of chemicals in food production is an issue that may influence consumer food choices. Concern about the use of chemicals in the production of food crops is probably greatest in relation to the use of insecticides (Endersby and Morgan, 1991). Genetic engineering of insect resistance offers an option for reduction in the use of chemicals in agriculture. Plant insect resistance is also a potentially valuable component of any integrated pest management strategy. Transgene options available for insect resistance include protease inhibitors, α-amylase inhibitors, lectins and bacterial toxins (Gatehouse and Hilder, 1994). The first three of these are common in plants, especially in seeds.

Protease inhibitors are able to prevent digestion of proteins by insects and hence slow the growth rate of the insect. These proteins probably function in the plant as natural protection against insect attack. A cowpea protease inhibitor (CpTI) was the first plant gene used successfully for the genetic engineering of insect resistance (Hilder *et al.*, 1987). The CpTI is a member of a class of double-headed serine protease inhibitors with low mammalian toxicity and broad insecticidal action.

Lectins are carbohydrate-binding proteins that may bind to the epithelia of the gut and be toxic to insects. α-Amylase inhibitors may also be effective in the control of insects.

Bacillus thuringiensis (Bt) produces crystalline proteins with insecticidal action against many Lepidopteran, Dipteran and Coleopteran

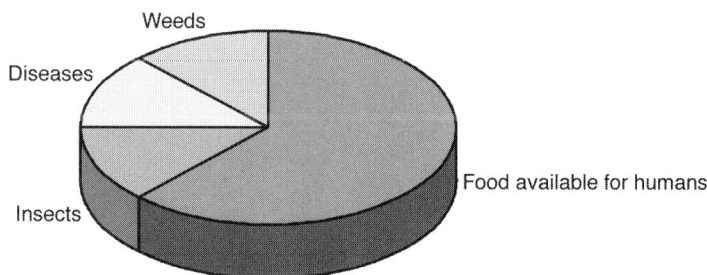

Figure 4.10 Pests reduce the available crops for human consumption. Transgenic plants may help reduce these losses.

insects. The intracellular crystals of Bt are produced by the bacteria during sporulation and, when ingested by some insects, are degraded in the insect gut by specific proteases. Toxic fragments are then released which disrupt the lining of the insect's gut.

Expression of Bt in the chloroplast has been shown to allow its production at 3–5% of total soluble protein (McBride *et al.*, 1995). A vector with Bt flanked by chloroplast sequences was used to achieve homologous recombination. This approach should allow very effective control of insects and may have value when other proteins are required at high levels in the plant.

Combinations of these genes may be used to ensure more durable resistance. Sweet potato (*Ipomoea batatas* (L.) Lam.) has been transformed to express both the cowpea trypsin inhibitor and snowdrop lectin (Newell *et al.*, 1995).

4.5.2 DISEASE RESISTANCE

Resistance to diseases caused by bacteria, fungi viruses and nematodes may be controlled by genetic engineering. Native disease resistance genes have been cloned by a variety of methods, transposon tagging, map-based cloning, and by homology. Novel proteins may be identified that confer resistance against more than one of these classes of plant pathogens.

(a) Bacteria and fungi

Proteins effective against bacteria may be isolated from many sources including insects and plants. Candidate genes may be identified by screening likely antimicrobial agents against cultures of the bacteria. Genes effective against fungi may be identified by a similar screening method to that used for bacteria. Constitutive expression of chitinases has been used to provide general fungal resistance. Chitinase probably acts by degrading fungal cell walls. Proteins that cause leakage of membranes such as thionins and permatins, including osmotin and zeamatin, have been shown to be effective against bacteria or fungi (Ward *et al.*, 1994). Blocking of ethylene production to control ripening may also result in increased resistance to bacterial and fungal spoilage organisms.

(b) Viruses

The production of transgenic plants with resistance to viruses (Figure 4.11) was one of the first successful practical applications of plant transformation. Several strategies involving the expression of part of

Figure 4.11 Field-testing virus-resistant potatoes. The coat protein gene from the virus was used to produce plants resistant to virus.

the viral genome in the plant have proven effective (Buck, 1991). The expression of the coat protein gene from the virus has met with widespread success, but expression of other viral genes – including the viral replicase gene – has also been successful. Both sense and antisense expression of parts of the viral genome may be protective against viral infection.

(c) Nematodes

Nematodes cause serious reductions in plant productivity in many environments. Marker-assisted selection has been used in the development of nematode-resistant plants. Native nematode resistance genes have been characterized and may provide useful protection when used as transgenes. Novel genes for nematode resistance offer an alternative approach to the production of resistant plants. Transgenic plants provide an opportunity to develop plants with genetic resistance to these long-term plant pests and thus contribute to a reduced reliance on chemical nematocides in agriculture.

4.6 Producing herbicide-resistant plants

4.6.1 HERBICIDE RESISTANCE AS A CONSEQUENCE OF TRANSFORMATION TECHNIQUE

The use of herbicide resistance genes as selectable markers may result in transgenic plants which have herbicide resistance in addition to the genes that were the objective of the transformation. Strategies for removal of such marker genes have been proposed. However, the removal of herbicide resistance would in most cases delay production of useful transgenic plants excessively. The use of selectable marker genes for resistance to herbicides that are not used in agriculture may avoid concerns that the widespread presence of herbicide resistance genes will lead to undesirable increases in the use of herbicides in agriculture. The advantages of herbicide resistance genes as selectable marker genes include their broad effectiveness against a wide range of plants and their greater stability in solution compared with antibiotics

4.6.2 HERBICIDE-RESISTANT PLANTS IN AGRICULTURE

The risks associated with the introduction of herbicide resistance genes into crop and pasture species include the possibility that the genes will escape into weed populations, or that the transgenic plant itself may become a weed. The choice of herbicide is a very important consideration. Selecting herbicides that have unique modes of action may be an important strategy. Multiple herbicide resistances in weeds may develop rapidly in some systems when several classes or groups of herbicide act on the same molecular target. Transgenic plants with such resistance genes may encourage greater use of these herbicides and accelerate the appearance of weed populations with multiple herbicide resistances. The most suitable herbicides for use in agriculture are those with low mammalian toxicity and high activity against plants. Many such compounds have been developed by attacking metabolic pathways that are unique to plants. The pathways of biosynthesis of essential amino acids are good examples of plant-specific metabolism and have no parallel in mammalian systems. Herbicide resistance genes may protect by detoxifying the herbicide (converting it to an inactive form) or may provide an alternative pathway that is not inhibited by the herbicide (e.g. by encoding an enzyme not inhibited by the herbicide).

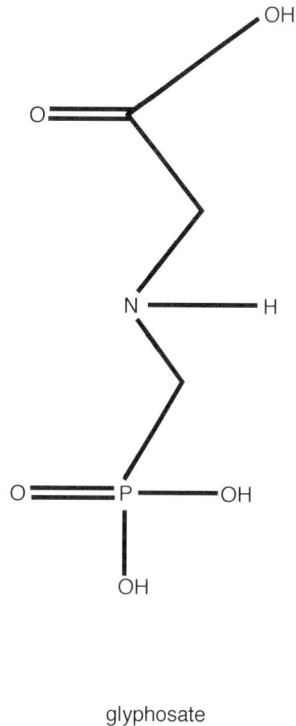

glyphosate

Figure 4.12 Structure of the herbicide, glyphosate.

(a) Glyphosate

Glyphosate (Figure 4.12) is a broad-spectrum herbicide used to control weeds in a variety of situations including agricultural and horticultural situations for food and ornamental crops; it is also widely used in private gardens. Glyphosate prevents the synthesis of aromatic amino acids in plants by inhibiting 5-enolypyruvyl shikimate-3-phosphate (EPSP). Several approaches to the engineering of glyphosate resistance have proven successful. Mutant EPSP genes have been recovered from *Salmonella typhimurium*, and overproduction of normal EPSP has conferred resistance.

(b) Sulphonylureas, imidazolinones and triazalopyrimidines

The sulphonylurea herbicides include chlorsulfuron and metsulfuron methyl, and demonstrate selectivity in cereal crops. Some of these herbicides are persistent in soil causing difficulties in crop rotations and the use of sulphonylurea herbicides has resulted in the appearance of resistant weed populations. These herbicides inhibit acetolactate synthase (ALS) and block the synthesis of the amino acids, leucine, isoleucine and valine. The imidazolinone and triazalopyrimidine classes of herbicide also act on ALS and cross-resistance to all three classes of herbicides is possible. Mutant ALS genes can confer resistance.

(c) 2-Dichlorophenoxyacetic acid (2,4-D)

The group of herbicides with auxin-type action are relatively toxic to animals compared with more modern herbicides but may be used in specific situations. A resistance gene encoding a protein degrading 2,4-D has been used to produce resistant plants.

(d) Phosphinothricin

Phosphinothricin (PPT) also known as glufosinate ammonium (formulated as Basta®) is an inhibitor of glutamine synthetase (Figure 4.13), causing an accumulation of ammonia that results in the herbicidal effect. The antibiotic bialaphos is an amino acid derivative of PPT. Plants cleave the amino acids to release active PPT (Figure 4.14). The enzyme phosphinothricin-N-acetyltransferase (PAT) inactivates PPT by acetylation (Figure 4.15). The gene for PAT from *Streptomyces hygroscopicus* (the source of bialaphos) is known as the *bar* gene and has been widely used as a selectable marker gene in the production of transgenic plants (Figure 4.16).

(e) Atrazine and bromoxynil

Atrazine is a member of the triazine group of herbicides used for non-selective control of weeds. Triazines and bromoxynil inhibit electron transport in photosystem II. These herbicides are relatively toxic and may persist in soil. A mutant chloroplast gene (*psb*A) protects against atrazine and a bromoxynil-degrading protein encoded by a gene from a soil bacterium is effective in transgenic plants.

4.6.3 RESISTANCE TO PARASITIC PLANTS

Transgenic plants resistant to herbicides have been used to achieve resistance to parasitic flowering plants. Transgenic plants which degrade the herbicide (e.g. the *bar* gene that confers resistance to glufosinate) have been shown to be less effective than those that tolerate the herbicide by expressing an enzyme resistant to the herbicide (e.g. a gene encoding a modified acetolactate synthase (ALS) for resistance to chlorosulfuron) (Joel *et al.*, 1995). Much of the damage caused by these parasites may

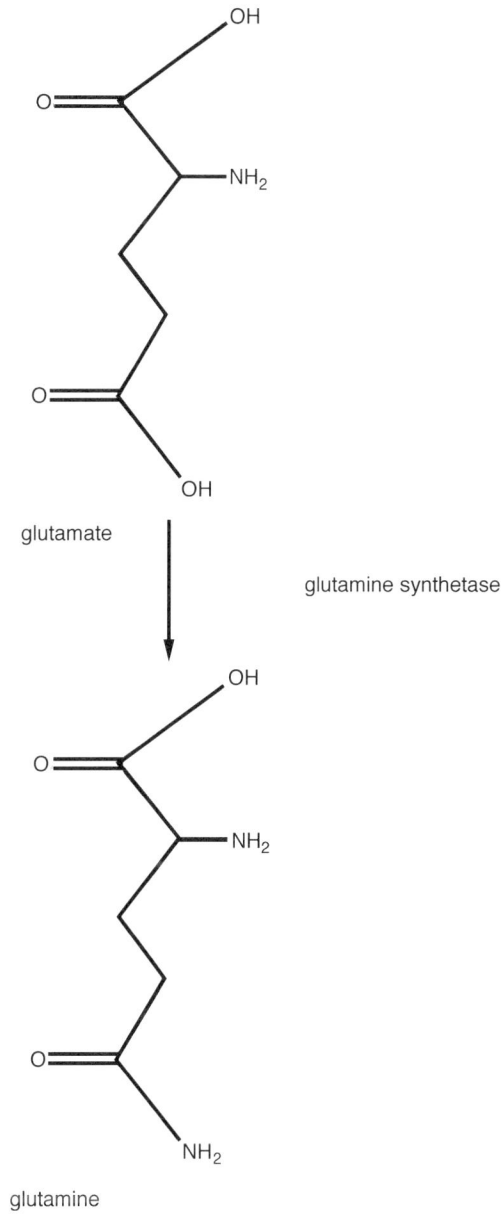

Figure 4.13 Glufosinate (an analogue of glutamate) inhibits glutamine synthetase.

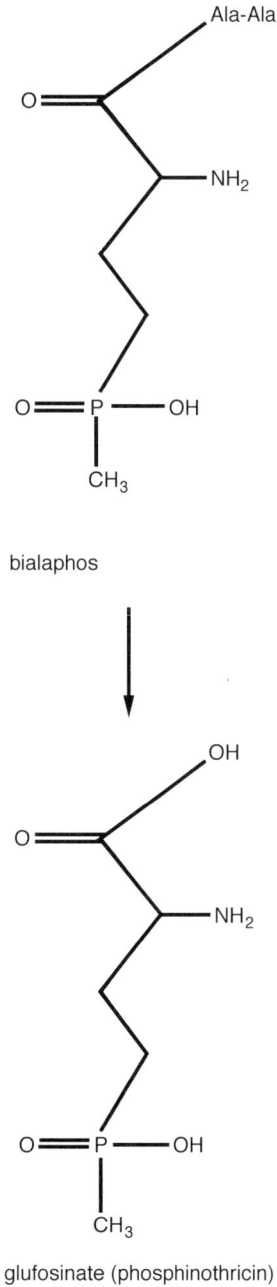

bialaphos

glufosinate (phosphinothricin)

Figure 4.14 Bialaphos is converted to phosphinothricin (PPT; Basta®) by removal of amino acid residues.

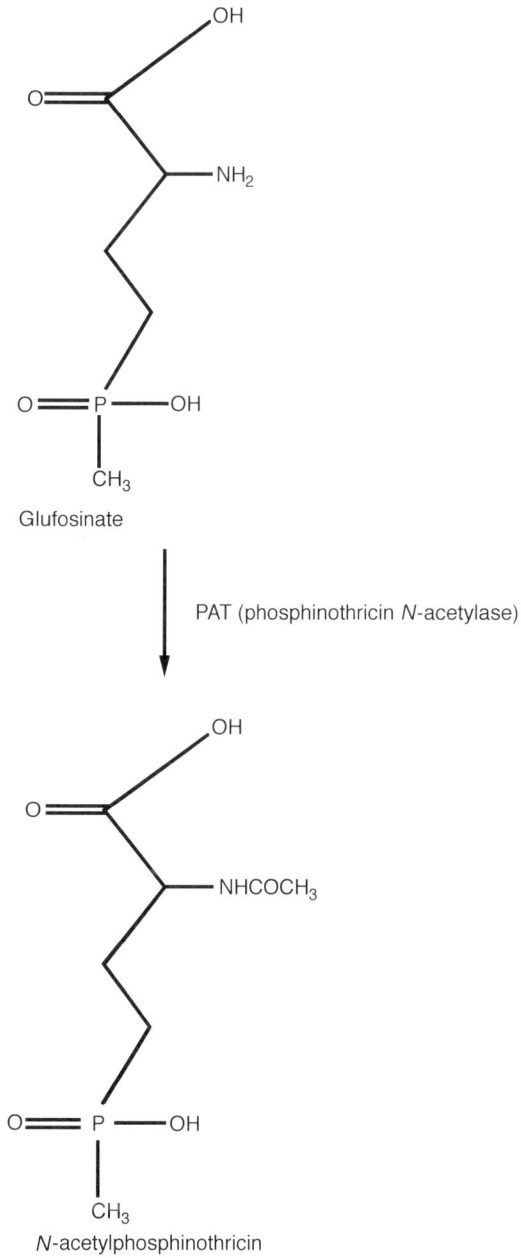

Figure 4.15 Phosphinothricin-*N*-acetylase (PAT) inactivates glufosinate (PPT) by acetylation.

Figure 4.16 Transgenic wheat plants resistant to the herbicide Basta® surrounding control plants. (From Kartha *et al.*, 1994.)

occur before the parasite emerges. The herbicide resistances based upon herbicide metabolism result in the herbicide being degraded before it reaches the root parasite, while the herbicide tolerance genes preserve the herbicide for action against the parasite.

4.7 Producing plants with resistance to abiotic stress

Transformation may also be used to develop plants with resistance to abiotic stress. Variations in water, nutrients, temperature and environmental contaminants may limit plant performance. Specific approaches have been proposed for each of these environmental constraints.

The production of transgenic plants with enhanced performance in relation to these stresses is likely to prove more difficult than other applications of plant transformation because of the genetic complexity of the processes involved. Rarely would a single gene be expected to be able consistently to enhance plant growth in response to these stresses. Exceptions might include genes allowing improved ability to osmoregulate in response to water stress, or genes allowing the complexing of heavy-metal contaminants in soil. Plants in the genus *Alyssum* (Brassicaceae) accumulate nickel at as much as 3% of dry weight, possibly because of their ability to complex the nickel with free histidine that these plants generate in large amounts (Kramer *et al.*, 1996). Transformation of plants to alter amino acid metabolism may offer an approach to the development of plants tolerant of soils contaminated with nickel residues. Such plants may have a role in the decontamination of polluted sites.

A gene encoding a protein from flounder has been used to protect plants against freezing damage. This protein may be useful in preventing frost damage in the field and also in protecting foods from freezing damage in storage after harvest. For example, it may be possible to use freezing to preserve the texture and flavour of some fruit or vegetables that are currently not suitable for freezing.

Transgenic plants with altered fertility may be generated by manipulating genes that control flowering and pollen formation. Expression of the *LEAFY* or *APETALA1* genes in *Arabidopsis* has been show to result in precocious flowering (Mandel and Yanofsky, 1995; Weigel and Nilsson, 1995). Early flowering may be useful in ornamental species. Genetically engineered male sterility has great potential for use in generation of hybrids in agriculture.

4.8 Manipulating reproductive behaviour in plants

Plants may be used to produce useful proteins (Owen, 1995; Whitelam, 1995). The advantage of plants for protein production may include low cost, the possibility of very large-scale production, and ease of recovery if the required protein is expressed in an appropriate part of the plant. Seeds may provide an ideal means for the storage and transport of recombinant proteins.

Industrial enzymes and antibodies are obvious candidates for production in plants. Proteins with application as pesticides have also been viewed as possible products from transgenic plants.

4.9 Producing recombinant proteins in plants

Plants have great potential as sources of antigens for the immunization of animals. Transgenic plants may be developed to produce antigenic proteins or other molecules. Production of the antigen in an edible part of the plant provides an effective delivery system for the antigen. Potential applications of this technology include efficient immunization of humans and animals against disease, and the control of animal pests. Antigens to the hepatitis B virus have successfully been expressed in tobacco plants and used to immunize mice (Thanavala *et al.*, 1995). Mice fed potatoes expressing the B subunit of the *Escherichia coli* enterotoxin LT-B produced antibodies protecting against the bacterial toxin. Diarrhoeal disease may cause up to 10 million human deaths per year, largely among children in the developing world (Blake and Arntzen, 1995). This suggests that inexpensive immunization against important human diseases may be possible by expression of antigens in the edible parts of plants. Oral vaccines against cholera have already been expressed in plants.

4.10 Using plants to deliver antigens

4.11 Consumer acceptance of transgenic plants

The ultimate success of plant genetic engineering depends upon consumer acceptance of the technology and the products generated. Consumers' concerns about the safety of genetic engineering may be balanced by their perceptions of the benefits and the risks of alternative technologies. For example, a transgenic pest-resistant plant that reduces reliance on the use of pesticides may be more acceptable than a transgenic herbicide-resistant plant that might be viewed as encouraging the use of more herbicide. Public perceptions of the risks associated with genetic engineering relative to those involved in the use of chemicals in agriculture will determine attitudes in these cases. Consumers may be more willing to accept products that benefit them directly rather than the producer. Products with improved quality are likely to meet less resistance than those that improve production efficiencies and are seen largely to benefit only the producer.

Acceptance is likely to be greater for plants that are not used for food, with transgenic cotton less likely to cause concern than transgenic wheat. Interest in the association between diet and health (e.g. the incidence of cardiovascular disease and cancer) is a major factor influencing consumer food choices. Transgenic food plants offering improved characteristics in this area may achieve greater market acceptance.

Consumers may be interested in both intrinsic and extrinsic factors when considering transgenic foods or other plant products. The intrinsic factors include the physical and chemical properties of the product itself, such as nutritional value, flavour, appearance, safety, and likely influence on health. Extrinsic factors including the method and ethics of production, influence on the environment and packaging may become increasingly important.

Involvement of consumers in public debates on the ethics of transgenic plants can be an important step towards gaining acceptance. Ultimately, although public attitudes are likely to determine the extent of application of molecular techniques (Sparks *et al.*, 1995), the way in which this technology is portrayed in the media is also likely to be influential (Frewer *et al.*, 1993).

4.12 Intellectual property issues

Transgenic plants may be of considerable commercial value. This value may be derived from the genes used or their method of introduction or deployment. It is possible for the rights to use a promoter, coding region and termination sequences in a useful gene construct to be owned separately. The transformation methods may also be subject to proprietary rights. This situation has generated considerable research, aiming to allow production of useful transgenic plants without infringing the patents of a competitor. The patenting of gene sequences (Yablonsky and Hone, 1995) may influence progress in the commercial exploitation of transgenic plants. The choice of gene sequences for use

in commercial transgenic plants may be driven primarily by considerations of ownership of intellectual property rather than technical
considerations. This may also determine the transformation protocol
employed.

The release of transgenic plants into the environment is regulated in
many countries to prevent dangerous products entering the food chain
or causing other harm to the environment. Release of plants with a
herbicide resistance gene may be considered unacceptable if the plant
is able to interbreed freely with closely related weed species (Mikkelsen
et al., 1996) allowing the development of herbicide-resistant weeds.
Plants producing foods with potentially toxic, allergenic or antinutritional properties may require careful evaluation. These different aspects
of risk evaluation are regulated by different agencies in some countries,
thus complicating the commercial release of transgenic plants. Special
containment facilities may be required for experiments involving transgenic plants (Figure 4.17).

| 4.13 Regulation of transgenic plants |

The production of transgenic plants without the presence of marker
genes (Yoden and Goldsbrough, 1994) may be necessary to satisfy
concerns about the safety of the products of these genes and plants
which express them. This either requires the development of efficient
transformation systems that do not require selectable marker genes,

Figure 4.17 Containment glasshouse for the growth of transgenic plants.

or their subsequent removal before release for commercial use. Although current selectable markers – largely antibiotic and herbicide resistance genes – may not be acceptable to consumers, other more acceptable selectable marker genes might be developed in the future.

The genetic engineering of reproductive sterility may also be a requirement for release of some transgenic plants into the environment (Strauss *et al.*, 1995). This may be essential to control the spread of genes into wild populations. The genetic engineering of male sterility has been demonstrated using a ribonuclease under the control of highly specific promoters.

The assessment of the competitive performance of transgenic plants and the associated risks of persistence of transgenic plants as weeds in the environment may require extensive field testing (Fredshaven *et al.*, 1995).

KEY TERMS

Agrobacterium
Direct DNA transfer
Electroporation
Gene silencing
Herbicide resistance
Homologous recombination
Intron
Marker gene
Methylation
Microprojectile bombardment
Promoter
Recombinant proteins
Regeneration
Selectable marker gene
Stable expression
Transcription
Transformation
Transient expression
Translation

EXAMPLES OF WORKED QUESTIONS

1. You have a herbicide resistance gene effective against herbicides that have value in the control of weeds in cereal crops. How would you approach the production of a herbicide resistant cereal crop?

   Microprojectile bombardment of immature embryos with the desired herbicide resistance gene construct followed by selection on media containing the herbicide and regeneration of plants from resistant cell lines.

2. What are the advantages of being able to target transgenes to specific locations on the chromosome?

   The number of copies of the gene introduced is controlled. The risk of loss of non-target gene function by insertional inactivation will be avoided. The level of expression will be defined by the site of insertion and it will not be necessary to screen large numbers of plants to identify those with an appropriate level of expression.

3. What steps could be taken to ensure that a transgenic food product was more readily acceptable to consumers?

   The use of selectable marker genes likely to cause poor acceptance could be avoided. The safety of the product could be demonstrated in animal feeding experiments. The equivalence of the food product to existing foods could be documented and promoted.

4. Outline the protocols used in transformation by microprojectile bombardment.

   A suitable gene construct is prepared and appropriate target tissues of the plant are generated. Gold or tungsten microparticles are coated with the DNA and accelerated into the plant tissue using a stream of inert gas (helium). Transgenic plants are regenerated from the bombarded tissue. This is usually achieved by including a selectable marker gene in the DNA preparation coated on the particles and growing the plant cells on an appropriate selection medium.

Questions

1. You have a gene encoding a protein that inhibits the growth of fungal pathogens of vegetable species. How would you approach the production of a vegetable crop with resistance to fungal disease?

2. List possible causes of failure to detect transgene expression in a plant in which the transgene has been detected by PCR.

3. How can intellectual property problems be minimized in the development of a transgenic plant product?

4. Outline the protocols used in *Agrobacterium*-mediated transformation.

References

Anderson, G.D., Blechl, A.E., Greene, F.C. and Weeks, J.T. (1994) Progress towards genetic engineering of wheat with improved quality, in *Improvement of Cereal Quality by Genetic Engineering* (eds R.J. Henry and J.A. Ronalds), Plenum Press, New York, pp. 87–95.

Baxter, D. (1995) The application of genetics to brewing. *Ferment*, 8, 307–15.

Birch, R.G. and Bower, R. (1994) Principles of gene transfer using particle bombardment, in *Particle Bombardment Technology for Gene Transfer* (eds N.-S. Yang and P. Christou), Oxford University Press, New York, pp. 3–37.

Blake, M.E. and Arntzen, C.J. (1995) Edible vaccines. *Probe*, 5, 1–2.

Buck, K.W. (1991) Virus-resistant plants, in *Plant Genetic Engineering* (ed. D. Grierson), Blackie, Glasgow, pp. 136–78.

Destefano-Beltran, L., Nagpala, P., Jaeho, K., Dodds, J.H. and Jaynes, J.M. (1991) Genetic transformation of potato to enhance nutritional value and confer disease resistance, in *Molecular Approaches to Crop Improvement* (eds E.S. Dennis and D.J. Llewellyn), Springer-Verlag, Vienna, New York, pp. 17–32.

D'Ovidio, R., Masci, S. and Porceddu, E. (1996) Sequence analysis of the 5' non-coding regions of active and inactive 1Ay HMW glutenin genes from wild and cultivated wheats. *Plant Science*, 114, 61–9.

Ebskamp, M.J.M., van der Meer, I., Spronk, B.A., Weisbeek, P.J. and Smeekens, S.C.M. (1994) Accumulation of fructose polymers in transgenic tobacco. *Bio/Technology*, 12, 272–5.

Endersby, N.M. and Morgan, W.C. (1991) Alternatives to synthetic chemical insecticides for use in crucifer crops, in *Biological Agriculture and Horticulture*, Vol. 8, A. B. Academic Publishers, UK, pp. 33–52.

Fredshaven, J.R., Poulsen, G.S., Huylbrechts, I. and Rudelsheim, P. (1995) Competitiveness of transgenic oilseed rape. *Transgenic Research*, 4, 142–8.

Frewer, L.J., Raats, M.M. and Shepherd, R. (1993) Modelling the media: the transmission of risk information in the British quality press. *IMA Journal of Mathematics Applied in Business and Industry*, 5, 235–47.

Gatehouse, A.M.R. and Hilder, V.A. (1994) Genetic manipulation of crops for insect resistance, In *Molecular Biology in Crop Protection* (eds G. Marshall and D. Walters), Chapman & Hall, London, pp. 177–201.

Gisel, A., Iglesias, V.A. and Sautter, C. (1996) Ballistic microtargeting of DNA and biologically active substances to plant tissue. *Plant Tissue Culture and Biotechnology*, **2**, 33–41.

Guo, Y., Liang, H. and Berns, M.W. (1995) Laser-mediated gene transfer in rice. *Physiologia Plantarum*, **93**, 19–24.

Henry, R.J. and Ronalds, J.A. (1994) *Improvement of Cereal Quality by Genetic Engineering*, Plenum Press, New York.

Hilder, V.A., Gatehouse, A.M.R., Sheerman, S.E., Barker, F. and Boulter, D. (1987) A novel mechanism of insect resistance engineered into tobacco. *Nature*, **330**, 160–3.

Holton, T.A. and Cornish, E.C. (1995) Genetics and biochemistry of antho-cyanin biosynthesis. *The Plant Cell*, **7**, 1071–83.

Joel, D.M., Kleifield, Y., Losner-Goshen and Herzlinger, G. (1995) Transgenic crops against parasites. *Nature*, **374**, 220–1.

Kartha, K.K., Nehra, N.S. and Chibbar, R.N. (1994) Genetic engineering of wheat and barley, in *Improvement of Cereal Quality by Genetic Engineering* (eds R.J. Henry and J.A. Ronalds), Plenum Press, New York, pp. 21–30.

Kramer, M.G. and Redenbaugh, K. (1994) Commercialization of a tomato with an antisense polygalacturonase gene: The FLAVR SAVR™ tomato story. *Euphytica*, **79**, 293–7.

Kramer, U., Cotter-Howells, J., Charnock, J.M., Baker, A.J.M. and Smith, J.A.C. (1996) Free histidine as a metal chelator in plants that accumulate nickel. *Nature*, **379**, 635–8.

Lambe, P., Dinant, M. and Matagne, R.F. (1995) Differential long-term expression and methylation of the hygromycin phosphotransferase (hph) and β-glucuronidase (GUS) genes in transgenic pearl millet (*Pennisetum glaucum*) callus. *Plant Science*, **108**, 51–62.

Li, Y., Ma, H., Zhang, J., Wang, Z. and Hong, M. (1995) Effects of the first intron of rice waxy gene on the expression of foreign genes in rice and tobacco protoplasts. *Plant Science*, **108**, 181–90.

Lumen, B.O. de (1990) Molecular approaches to improving the nutritional and functional properties of plant seeds as food sources: developments and comments. *Journal of Agricultural and Food Chemistry*, **38**, 1779–88.

MacGregor, A.W. (1996) Malting and brewing sciences: challenges and oppor-tunities. *Journal of the Institute of Brewing*, **102**, 97–102.

Mandel, M.A. and Yanofsky, M.F. (1995) A gene triggering flower formation in *Arabidopsis*. *Nature*, **377**, 522–4.

Matzke, M.A. and Matzke, A.J.M. (1995) How and why do plants inactivate homologous (trans)genes? *Plant Physiology*, **107**, 679–85.

McBride, K.E., Svab, Z., Schaaf, D.J., Hogan, P.S., Stalker, D.M. and Maliga, P. (1995) Amplification of a chimeric *Bacillus* gene in chloroplasts leads to an extraordinary level of an insecticidal protein in tobacco. *Bio/Technology*, **13**, 362–5.

McElroy, D. and Brettell, R.I.S. (1994) Foreign gene expression in transgenic cereals. *Trends in Biotechnology*, **12**, 62–8.

McElroy, D. and Jacobsen, J. (1995) What's brewing in barley biotechnology? *Bio/Technology*, **13**, 245–9.

McKinnon, G.E., Abedinia, M. and Henry, R.J. (1996) Expression of non selectable markers in wheat and rice tissues. *Plant Tissue Culture and Biotechnology*, **2**, 24–32.

Mikkelsen, T.R., Anderson, B. and Jorgensen, R.B. (1996) The risk of crop transgene spread. *Nature*, **380**, 31.

Morton, R. and Hooykaas, P.J.J. (1995) Gene replacement. *Molecular Breeding*, **1**, 123–32.

Nawrath, C., Poirier, Y. and Somerville, C. (1995) Plant polymers for biodegradable plastics: cellulose, starch and polyhydroxyalkanoates. *Molecular Breeding*, **1**, 105–22.

Newell, C.A., Lowe, J.M., Merryweather, A., Rooke, L.M. and Hamilton, W.D.O. (1995) Transformation of sweet potato (*Ipomoea batatas* (L.) Lam.) with *Agrobacterium tumefaciens* and regeneration of plants expressing cowpea trypsin inhibitor and snowdrop lectins. *Plant Science*, **107**, 215–27.

Offringa, R. and Hooykaas, P. (1995) Gene targeting in plants, in *Gene Targeting* (ed. M.A. Vega), CRC Press, Boca Raton, pp. 83–121.

Owen, M.R.L. (1995) The production of recombinant proteins in plants; University of Leicester, UK. 24–27 July, 1994. *Transgenic Research*, **4**, 151–2.

Romano, C.P. and Klee, H. (1993) Hormone manipulation in transgenic plants, in *Transgenic Plants, Fundamentals and Applications* (ed. A. Hiatt), Marcel Dekker, New York, pp. 23–36.

Sagi, L., Panis, B., Remy, S., Schoofs, H., De Smet, K., Swennen, R. and Cammue, P.A. (1995) Genetic transformation of banana and plantain (*Musa* spp.) via particle bombardment. *Bio/Technology*, **13**, 481–5.

Shewry, P.R., Tatham, A.S., Halford, N.G., Barker, J.H.A., Hannappel, U., Gallois, P., Thomas, M. and Kreis, M. (1994) Opportunities for manipulating the seed protein composition of wheat and barley in order to improve quality. *Transgenic Research*, **3**, 3–12.

Shewry, P.R., Tatham, A.S., Barro, F., Barcelo, P. and Lazzeri, P. (1995) Biotechnology of breadmaking: unraveling and manipulating the multiprotein gluten complex. *Bio/Technology*, **13**, 1185–90.

Smith, C.R., Saunders, J.A., Van Wert, S., Cheng, J. and Matthews, B.F. (1994) Expression of GUS and CAT activities using electrotransformed pollen. *Plant Science*, **104**, 49–58.

Sparks, P., Shepherd, R. and Frewer, L.J. (1995) Assessing and structuring attitudes towards the use of gene technology in food production: the role of perceived ethical obligation. *Basic and Applied Social Psychology*, **16**, 267–85.

Stark, D.M., Timmerman, K.P., Barry, G.F., Preiss, J. and Kishore, G.M. (1992) Regulation of the amount of starch in plant tissues by ADP glucose pyrophosphorylase. *Science*, **258**, 287–92.

Strauss, S.H., Rottermann, W.H., Brunner, A.M. and Sheppard, L.A. (1995) Genetic engineering of reproductive sterility in forest trees. *Molecular Breeding*, **1**, 5–26.

Thanavala, Y., Yang, Y.-F., Lyons, P., Mason, H.S. and Arntzen, C. (1995) Immunogenicity of transgenic plant-derived hepatitis B surface antigen. *Proceedings of the National Academy of Sciences, USA*, **92**, 3358–61.

Vickers, J.E., Hamilton, S.E., de Jersey, J., Henry, R.J., Marschke, R.J. and

Inkerman, P.A. (1996) Assessment of *Bacillus licheniformis* α-amylase as a candidate for genetic engineering of malting barley. *Journal of the Institute of Brewing*, **102**, 75–8.

Ward, E., Uknes, S. and Ryals, J. (1994) Engineering resistance to diseases, herbicides and pests, in *Molecular Biology in Crop Protection* (eds G. Marshall and D. Walters), Chapman & Hall, London, pp. 119–45.

Weigel, D. and Nilsson, O. (1995) A developmental switch sufficient for flower initiation in diverse plants. *Nature*, **377**, 495–500.

Whitelam, G.C. (1995) The production of recombinant proteins in plants. *Journal of the Science of Food and Agriculture*, **68**, 1–9.

Yablonsky, M.D. and Hone, W.J. (1995) Patenting DNA sequences. *Bio/Technology*, **13**, 656–7.

Yoder, J.I. and Goldsbrough, A.P. (1994) Transformation systems for generating marker-free transgenic plants. *Bio/Technology*, **12**, 263–7.

Useful routine protocols in plant molecular biology | 5

Efficient and reproducible protocols for sample handling and manipulation are crucial in the practical application of plant molecular biology in the laboratory.

Chapter outline

Practical application of plant molecular biology requires the choice of
protocols appropriate to the objective of the work. The literature
contains large numbers of procedures, many of which are variations
on a theme. Often a technique is favoured because of familiarity or
successful personal experience, while others that may be more suited
to the task are not considered.

The development of techniques for use in biological research is often
driven by the desire to simplify procedures to reduce the risk of error.
Complex procedures may be more difficult to reproduce. The philos-
ophy of method development demands that a protocol should contain
only the minimum number of steps required to achieve a reliable result.
Indeed, additional unnecessary steps are viewed as risking the introduction of variability and increasing the difficulty of reproducing
results. Plant molecular biology has been subject to this scrutiny to
varying degrees in different areas and great simplification of many
current protocols may be possible with further study.

This chapter brings together protocols that have been found to be
of use in practical applications of plant molecular biology. The
emphasis is on simple and reliable approaches with some background
on the basis for choice of specific methods in particular circumstances.

5.1 Introduction

5.2.1 INTRODUCTION

DNA isolation methods vary enormously, and before beginning any
study of plant DNA the objectives of the study must be defined so as
to establish the most appropriate sampling and DNA isolation
protocol. The quantity and quality of DNA required will define the

**5.2 Isolation of
plant DNA**

best option. For example, DNA isolated to prepare genomic libraries would be needed in larger amounts and of a higher purity than would DNA to be used in a screening method of a plant breeding programme based on a robust PCR.

5.2.2 SAMPLE COLLECTION AND STORAGE

Fresh plant material is highly suitable for isolation of DNA but is not always available. Frozen material can generally be stored at −20°C for short periods, but depending upon the quality of the DNA required may need to be stored at −70°C or a lower temperature (e.g. liquid nitrogen). Care must be taken not to allow frozen material to thaw for even a brief period. Freezing also preserves nuclease activity and degradation of DNA in thawed tissue may be rapid because of subcellular disruption caused by the freezing. Fresh material may sometimes be kept for several days in a cold room or refrigerator at around 4°C without harming the subsequent DNA preparation. Dry plant material may also be a good source of DNA, allowing extraction from herbarium samples and permitting the collection and storage of large numbers of samples at room temperature at low cost (Thomson and Henry, 1993). The method of drying is important and slow drying under conditions that maintain the integrity of the tissue seems to give the best results. Leaf samples collected in the field and placed in paper bags have been found to yield good quality DNA when analysed after more than a year of storage in an air-conditioned laboratory. A sap extractor (Clarke et al., 1989) has been designed for the preparation of DNA from large numbers of fresh plant samples for analysis by techniques requiring relatively large amounts of DNA (e.g. RFLP). Tissues such as flower petals may be a good source of DNA in some species (Lin and Ritland, 1995).

5.2.3 STANDARD PROCEDURES FOR GENERAL USE

The following procedure (Graham et al., 1994) has been found to be of general use and can be scaled up or down for different applications.

Plant DNA isolation

- Grind sample to a fine powder in liquid nitrogen using a mortar and pestle.
- Transfer the tissue to a tube of suitable size and add 1 ml of extraction buffer (2% (w/v) hexadecyltrimethylammonium bromide

Herbarium sample

DNA may be extracted from herbarium samples for analysis by PCR.

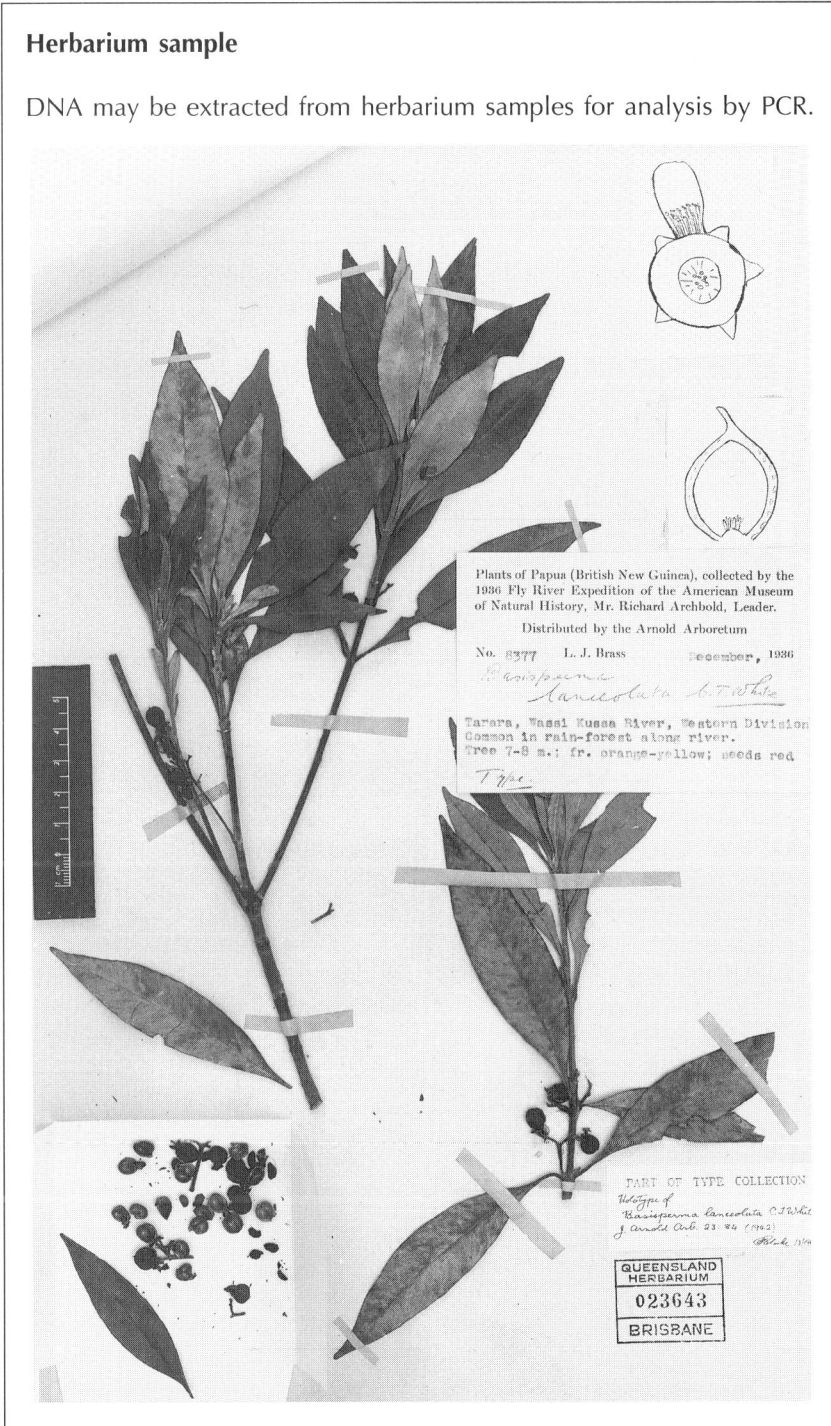

Plants of Papua (British New Guinea), collected by the
1936 Fly River Expedition of the American Museum
of Natural History, Mr. Richard Archbold, Leader.

Distributed by the Arnold Arboretum

No. 8377   L. J. Brass   December, 1936

Tarara, Wassi Kussa River, Western Division
Common in rain-forest along river.
Tree 7-8 m.: fr. orange-yellow; seeds red

PART OF TYPE COLLECTION

QUEENSLAND
HERBARIUM
023643
BRISBANE

(CTAB), 100 mM Tris–HCl, 1.4 M NaCl, 20 mM EDTA) per gram of tissue.
- Mix by gentle inversion and heat at 55°C for 20 min.
- Centrifuge at $15\,000 \times g$ for 5 min.
- Transfer the supernatant to a fresh tube and add one volume of chloroform : isoamyl alcohol (24 : 1)
- Mix by gentle inversion for 2 min and centrifuge at $15000 \times g$ for 20 s.
- Transfer the upper aqueous phase to a fresh tube and add 1/10 volume of 7.5 M ammonium acetate and two volumes of ice-cold ethanol.
- Mix by gentle inversion and place in a freezer at –20°C for 60 min.
- Centrifuge at $1500 \times g$ for 1 min and discard the supernatant.
- Wash the pellet twice with 70% (v/v) ethanol, mixing each time by gentle inversion.
- Dry the DNA in a desiccator.
- The DNA may be resuspended in an appropriate volume of TE buffer (10 mM Tris–HCl, 1 mM EDTA, pH 8.0) or solution appropriate for the application intended.

5.2.4 SIMPLE PROCEDURES FOR USE IN PCR SCREENING

The most extreme approach that can be taken is to avoid the DNA isolation step completely and introduce the plant material directly into the PCR reaction. This may require some modification or optimization of the PCR protocol. Plastic pipette tips may be used to punch a sample from leaves or roots of some plants (Berthomieu and Meyer, 1991). Tougher tissues may require the development of specialized tools. Others have found that pre-treatment of the plant tissue is helpful before PCR. Klimyak et al. (1993) heated in alkali before PCR. The sample was heated in 40 µl of 0.25 M NaOH on a boiling water bath for 30 s, neutralized by adding 40 µl of 0.25 M HCl and 20 µl of 0.5 M Tris–HCl, pH 8.0, 0.25% (v/v) Nonidet P-40, before boiling for a further 2 min. The time of heating may need to be established empirically for each tissue type.

Direct introduction of plant tissue with or without pre-treatment may not prove successful with all samples because of the presence of inhibitory substances. It is also more difficult with PCR in small volumes because of the need to introduce such a small piece of tissue. More general simple protocols for extracting DNA have been developed for this purpose. Steiner et al. (1995) reported a rapid one-step extraction buffer (ROSE) containing 10 mM Tris–HCl, pH 8.0, 312.5 mM EDTA, 1% sodium lauryl sarkosyl, and 1% polyvinylpolypyrrolidone. Ground lyophilized tissue was heated in this buffer at 90°C for 20 min. The DNA may be extracted by heating in a salt solution (Thomson and Henry, 1995). Seeds have been analysed by PCR following extraction of DNA from half the seed (Chunwongse *et al.*, 1993). In this procedure, the half-seeds were heated at 50°C for 1 hour in 200 µl of 10 mM Tris–HCl, pH 8.0, 50 mM KCl, 1.5 mM MgCl$_2$, 0.01% gelatin, 0.45% NP-40, 0.45% Tween-20 and 12–48 µg proteinase K. The seeds were then heated at 100°C for 5 min and the supernatant used directly in PCR.

Squashing the plant sample onto a nylon membrane is a way of processing large numbers of samples (Langridge *et al.*, 1991). This also provides a convenient method for storing large numbers of samples before PCR analysis.

The following procedure has been found to be of wide general use in preparing plant material for PCR analysis.

DNA preparation for PCR

- Place a small piece (0.5–1 mg works well for most tissues) of the plant tissue in a microfuge tube.
- Add 20 µl of 100 mM Tris–HCl, pH 9.5, 1 M KCl, 10 mM EDTA.
- Heat at 95°C for 10 min.
- Add 1 µl directly to the PCR (a dilution 1 : 10 on 1 : 100 may improve results).

The preparations should be stable for months when stored at room temperature. Some preparations may need to be diluted with water or PCR buffer for use in certain PCR systems.

5.2.5 REMOVAL OF SPECIFIC CONTAMINANTS

Another approach to preparing DNA from difficult tissues is specifically to target the removal of contaminants that are interfering with the subsequent use or analysis of the DNA. Removal of contaminating polysaccharides using hydrolytic enzymes (a mixture of glycoside hydrolases) has been used to isolate DNA free from polysaccharides (Rether *et al.*, 1993). Ion exchangers have been used to separate DNA from contaminating polysaccharides and other components (Marechal-Drouard and Guillemaut, 1995). A study of polysaccharide inhibition of RAPDs concluded that dilution of DNA preparation was the best way to reduce the problem (Pandey *et al.*, 1996)

SAMPLE	A320	A280	A260	280/260	260/280	PROTEIN	NUCLEIC ACID
1.0000	0.0050	0.0038	0.0040	1.2000	0.8333	-1.105	-0.020
2.0000	-0.001	0.0530	0.1027	0.5212	1.9187	5.3562	4.5832
3.0000	-0.003	0.0490	0.0958	0.5259	1.9015	5.8207	4.3440
4.0000	-0.003	0.0778	0.1511	0.5245	1.9065	8.7486	6.7839
5.0000	0.0011	0.0833	0.1572	0.5269	1.8979	9.4371	6.8568

Figure 5.1 Printout of UV analysis of plant DNA preparation. The A_{260} allows estimation of concentration and the A_{260}/A_{280} is a measure of purity.

5.2.6 ESTIMATION OF THE QUANTITY OF DNA OBTAINED

Several options may be considered for the estimation of the concentration of DNA present in an extract. The concentration can be estimated from the A_{260} if the preparation is relatively pure. The protein content can be deduced from the A_{280}. An A_{260}/A_{280} ratio above 1.8 indicates pure DNA (Figure 5.1). Crude preparations may contain too much other material absorbing in the UV that will interfere with this determination. Fluorimetic tests based upon simple dedicated fluorometers are widely available. These involve the addition of a fluorochrome such as H33258 (bisbenzimide) (Hoechst). When excited with light at 360 nm, the fluorescence of the dye at 458 nm increases in the presence of double-stranded DNA. The dye binds to specific sequences (two or three consecutive AT base pairs) and may thus be influenced by the base composition and sequence of the DNA and is therefore not usually suitable for estimation of concentrations of short DNA molecules (< 200 bp). This method does have the advantage of avoiding the interference of RNA that is a major problem when using UV absorbance to measure DNA concentration. Analysis by agarose gel electrophoresis is probably the most reliable approach, especially for crude preparations. An aliquot of the sample can be briefly run on an agarose gel, stained with ethidium bromide and compared with a known amount of a marker run in a lane on the same gel. This will permit a rapid visual estimate of the approximate amount of DNA in the extract. The advantage of this approach is that other nucleic acid, degraded DNA and other substances that interfere with UV or fluorimetic analysis may be separated from the high-molecular weight genomic DNA, allowing a more reliable estimate of yield. Most applications do not require that the amount of DNA be known exactly. Techniques such as RAPD analysis depend more upon ensuring a consistent amount of DNA from sample to sample.

5.2.7 ISOLATION OF DNA FROM MITOCHONDRIA AND
CHLOROPLASTS

The specific isolation of DNA from organelles is best achieved by first
separating the organelles, usually by sucrose gradient centrifugation.
The organelles can then be washed to remove contaminating DNA
before lysis to release the organellar DNA. A procedure for the isolation
of chloroplast DNA is described by Michaud *et al.* (1995).

5.3.1 SOUTHERN

Southern blotting (see Box) is used to analyse DNA for the presence
of specific sequences. The DNA is usually separated by gel electrophoresis and then blotted onto a membrane to allow hybridization
with a specific labelled DNA (or RNA) probe (Figure 5.2). Southern
blotting is an integral part of the RPLP marker technique and is an
important technique for the characterization of transgenic plants, being
used to establish that genes have been integrated into the genome and
to determine copy number.

5.3 Blotting

(a) RFLP analysis

RFLP analysis of plant genomic DNA involves isolation of the DNA,
digestion with a restriction enzyme, separation by agarose gel electrophoresis, transfer to a membrane, hybridization with a labelled probe,
and detection. The protocols for DNA isolation described above are
suitable for use in RFLP analysis. The other protocols required will
be outlined below.

(b) Digestion with restriction enzymes

Genomic Southerns require the isolation and digestion of larger
amounts of DNA from plants with a larger genome. For example,
analysis of 5 μg per lane was found to be sufficient for soybean (Apuya
et al., 1988). The sample should be digested in the buffer recom-
mended by the enzyme supplier. The digestion was conducted overnight
(12–16 h) with 5 units of enzyme per μg of DNA in the soybean case
described above. The objective is to ensure that the digestion has
proceeded to completion.

(c) Transfer to nylon membranes

The usual approach is to transfer to nylon membranes by capillary
blotting. Electrophoretic or vacuum blotting is also possible.

Southern blotting

- After running the agarose gel (see p. 190), cut the top right-hand corner to allow correct orientation in subsequent steps. Photograph the gel with the molecular weight markers showing and with a scale in millimetres (a plastic ruler with clear markings placed down the side of the gel) to allow subsequent estimation of the size of fragments on the membrane.
- Depurinate by transferring the gel to a solution of 0.25 M HCl and agitate by gentle rocking for 10 minutes.
- Rinse briefly in distilled water.
- Denature by gentle shaking in 0.4 M NaOH, 1 M NaCl for 15 minutes.
- Transfer DNA to a nylon membrane by capillary action. Place gel on top of 3MM paper on a support in a tray of $10 \times SSC$ (1.5 M NaCl, 2 M sodium citrate, pH 7.0); carefully place nylon membrane (wetted with $2 \times SSC$) on gel. Care should be taken to avoid air bubbles being trapped between the gel and the membrane. Cover with two layers of 3MM paper (wetted with $2 \times SSC$) and place a stack of dry paper towels on top. Allow transfer to proceed for 8–24 hours.
- Remove membrane from gel. The positions of the lanes and the edges of the gel can be marked on the membrane with a pencil before removing the membrane.
- Neutralize the membrane by soaking in 0.5 M Tris–HCl, pH 7.2, 1 M NaCl for 15 minutes.
- Fix DNA to membrane by drying the membrane briefly by blotting with 3MM paper and either baking in an oven for 2 hours at 80°C or exposing to UV light (254 nm). The membrane may be wrapped in plastic film and placed face-down on a UV transilluminator for 3–4 minutes. **Caution**: take care to avoid self-exposure to UV radiation.

(d) Labelling of probes

Several options are available for labelling probes. Radioactive labels offer good sensitivity but non-radioactive labels may be preferred because of greater stability of the labelled probe and ease of handling and disposal.

Protocols available for incorporation of the label include nick translation, random primer methods and PCR.

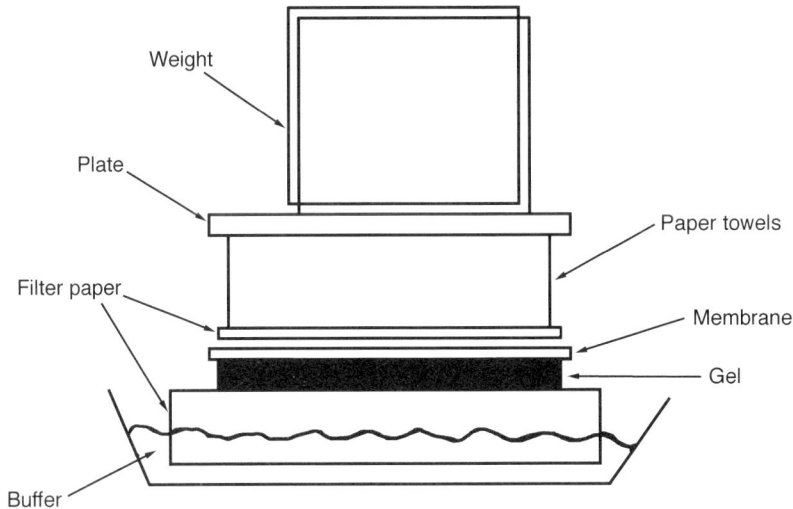

Figure 5.2 Southern blotting. A schematic representation of the blotting technique.

Labelling of probes with digoxigenin (DIG)

Labelling by PCR
Digoxigenin-11–dUTP (DIG–dUTP) may be incorporated by the action of *Taq*DNA polymerase in a polymerase chain reaction. The procedure involves the addition of DIG–dUTP to the PCR reaction mix to substitute for about one-third of the dTTP.

Labelling by random priming
(This procedure is as described by Boehringer-Mannheim in the DIG User's Guide.)

- Heat 1 µg of the DNA to be labelled (10 min in a boiling water bath) in 15 µl of water to denature, and quickly chill on ice.
- Add 2 µl of a mixture of random hexanucleotides (1.6 mg/ml in 500 mM Tris–Hcl, 100 mM $MgCl_2$, 1 mM dithiothreitol, 2 mg/ml BSA, pH 7.2).
- Add 2 µl of a solution containing 1 mM dATP, 1 mM dCTP, 1 mM dGTP, 0.65 mM dTTP and 0.35 mM alkali-labile DIG–dUTP, pH 6.5.
- Add 1 µl Klenow enzyme to give a final concentration of 100 U/ml.
- Incubate at 37°C for 1 h or more.
- Add 2 µl EDTA to terminate the reaction.

(e) Hybridization

The labelled probe is hybridized to the membrane under conditions of appropriate stringency. At high salt and low temperatures non-specific hybridization is possible. The stringency is increased by increasing the temperature and lowering the salt concentration.

Hybridization with labelled probe

- Prehybridize by incubating membrane in 10 ml of pre-hybridization solution at 50°C for 30–60 min.
- Denature probe at 100°C for 10 min and quench on ice.
- Hybridize overnight (either in a sealed plastic bag or a hybridization oven).
- Wash membrane three times in $2 \times SSC$, 0.1% SDS at room temperature.
- Wash membrane twice in $0.2 \times SSC$, 0.1% SDS at required temperature (e.g. 68°C).
- Remove membranes and continue to detection, or dry and store between filter paper at room temperature.

(f) Detection

The detection method will depend upon the type of label used. A digoxigenin (DIG)-labelled probe may be detected colorimetrically or using chemiluminescent methods.

Colorimetric detection

(This method uses components of the DIG DNA Labelling and Detection Kit, Boehringer-Mannheim.)

- Equilibrate in washing buffer (maleic acid buffer) for 1–3 min.
- Block by gently agitating in 20 ml of blocking solution (2% blocking reagent in washing buffer) for 30–60 min.
- Add 4 μl of anti-DIG alkaline phosphatase antibody the solution and incubate for 30 min.
- Remove antibody solution and wash membrane twice in washing buffer for 15 min.
- Prepare colour solution by mixing 45 μl of NBT solution (75 mg/ml nitroblue tetrazolium in 70% dimethylformamide) and 35 μl of X-Phosphate solution (50 mg/ml 5-bromo-4-chloro-3-indolyl phosphate in dimethylformamide) in 10 ml of 1 M Tris–HCl, pH 9.5, 1 M NaCl.
- Incubate the membrane in the colour solution in the dark. Do not shake.

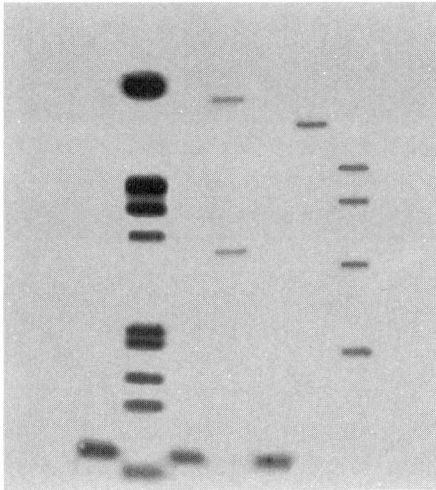

Figure 5.3 Southern blot example. Screening of transgenic rice plants with a DIG-labelled probe for a hygromycin resistance gene.

- Stop colour development by transferring the membrane to Tris–EDTA (TE) buffer.

Chemiluminescent detection

- Process as for colorimetric detection until after antibody binding and washing.
- Equilibrate in 10 ml of detection solution (0.1 M Tris–HCl, pH 9.5, 0.1 M NaCl) for 2–5 min.
- Remove membrane and place face-up on a sheet of plastic film. Cover membrane with substrate solution (10 µl of 10 mg/ml CSPD* substrate in 1 ml of detection buffer) and incubate for 5 min (1 ml will be sufficient for about 25 cm² of membrane).
- Remove excess liquid by blotting with filter paper and incubate between sheets of overhead transparency film for 10 min at 37°C.
- Expose to X-ray film for 1–4 hours and develop film.
* CSPD = disodium 3-(4-methoxyspiro{1,2-dioxethane-3,2'-(5'-chloro)tricyclo[3,3,1,1]decan}-4-yl)phenyl phosphate].

   Southern blots (Figure 5.3) provide key information on transgenic plants.

5.3.2 NORTHERN

Northern blotting is used to analyse RNA for the presence of specific sequences in a method analogous to the Southern blotting procedure used with DNA. Northern blotting is an important technique in the analysis of gene expression. Analysis of the tissue and developmental patterns of mRNA expression can be monitored by Northern blotting if a probe for the gene of interest is available.

5.3.3 WESTERN

Western blotting (see Box) is used to analyse for the presence of a specific protein. The proteins are separated by electrophoresis and blotted onto a membrane to allow probing with specific, labelled anti-bodies. This technique provides an important approach to analysis of the expression of foreign proteins in transgenic plants.

5.4 PCR protocols

5.4.1 INTRODUCTION

The polymerase chain reaction (PCR) is one the most useful and widely used protocols in applied plant molecular biology. Many molecular marker methods are based upon PCR. Some of these are anonymous, while others are from known genes. PCR is also widely applied in the cloning of plant genes in analyses ranging from determination of sequence differences in phylogenetic studies to the production of useful gene constructs for plant transformation. Some useful PCR protocols are described here. Most such protocols that are to be repeated more than a few times are best conducted by preparing a bulk mixture of all the components that do not need to be varied. This 'master mix' can improve the reproducibility and reliability of results by reducing pipetting errors. The master mix can contain all components except the template. PCR can then be performed by a single addition of this solution to the template followed by temperature cycling. This reduces the variations that can be introduced by pipetting each of the components separately.

Non-specific priming of PCR can be reduced by using a 'hot start'. In this approach the PCR mixture is only completed after the solution has reached a temperature that ensures more specific primer annealing. This can be achieved by adding one of the components (e.g. the DNA polymerase) after the solution has been heated.

Western blotting

The example given is the bifunctional α-amylase/subtilisin inhibitor (BASI) from barley. The illustration shows:

(a) nitrocellulose membrane stained with amido black
(b) nitrocellulose membrane probed with a monoclonal antibody. Lane 1, markers; lane 2, purified protein; lane 3, extract from barley; lane 4, extract from wheat. The antibody was raised against the barley protein but also detects the related wheat protein (Jarrett *et al.*, 1996).

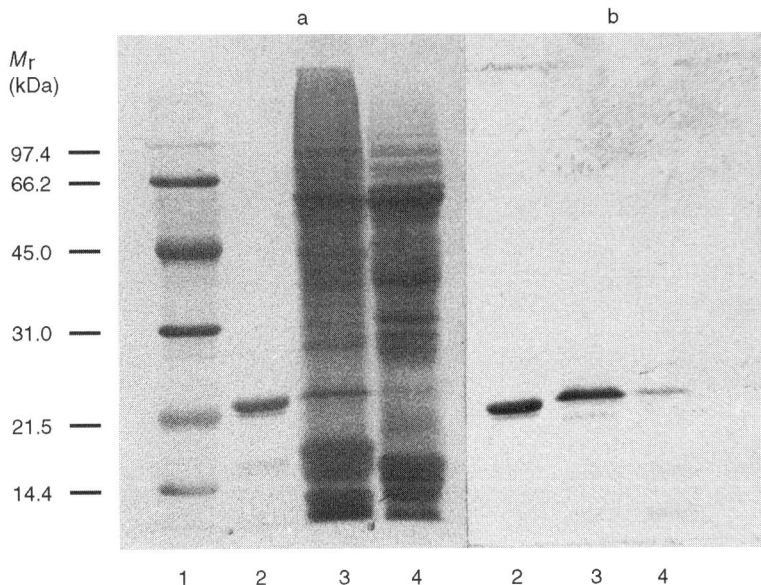

5.4.2 ARBITRARY PRIMER PROTOCOLS

Optimization of the DNA extraction and PCR steps may be necessary to achieve reproducible results with methods based upon PCR with short primers of arbitrary sequence (Yu and Pauls, 1994). The DNA extraction method needs to produce a consistent template for PCR amplification but this does not always require a complex extraction procedure. The same procedure will not be successful with all plant materials but the following protocols have been found to be reliable with a wide range of plants.

(a) RAPD

(This method is adapted from Williams *et al.*, 1990.)

- Prepare a DNA template using one of the methods described above, and dilute to 10 ng/μl.
- Add the following to a tube suitable for use in PCR:

    12.5 μl distilled water

    2.5 μl deoxynucleoside triphosphates (stock solution 1 mM to give a final concentration of 100 μM for each dNTP

    2 μl MgCl$_2$ (25 mM to give a final concentration of 2 mM)

    2.5 μl reaction buffer (100 mM Tris–HCl, 500 mM KCl, pH 8.3, to give a final concentration of 10 mM Tris–HCl and 50 mM KCl)

    0.5 μl thermostable DNA polymerase [*Taq* DNA polymerase (Perkin-Elmer, Cetus) to give a final concentration of 0.02 U/μl]

    2.5 μl of primer (2 μM to give a final concentration of 1 ng/μl)

- Centrifuge briefly to mix the contents of the tube.
- Add mineral oil (if necessary for thermal cycler).
- Cycle at 94°C for 1 min, 36°C for 1 min, and 72°C for 2 min; 45 cycles.
- Analyse by electrophoresis on agarose gels (described below) with detection by ethidium bromide staining (described below).

The same effect can be obtained by using a specially modified DNA polymerase (AmpliTaq Gold, Perkin-Elmer) that is not active at room temperature (Birch *et al.*, 1996). The enzyme becomes active only after heating.

(b) DNA Amplification Fingerprinting (DAF)

(This method is adapted from Caetano-Anolles and Bassam, 1993.)

- Prepare a DNA template using one of the methods described above and dilute to 1–10 ng/μl.
- Add the following to a tube suitable for use in PCR:

    12.5 μl distilled water

    2.5 μl deoxynucleoside triphosphates (stock solution 2 mM to give a final concentration of 200 mM for each dNTP)

    1.5 μl MgCl$_2$ (25 mM or 100 mM to give a final concentration of 1.5 or 6 mM)

    2.5 μl reaction buffer (100 mM Tris–HCl, 100 mM KCl, pH 8.3, to give a final concentration of 10 mM Tris–HCl and 10 mM KCl)

    0.75 μl thermostable DNA polymerase [AmpliTaq Stoffel fragment (Perkin-Elmer, Cetus) to give a final concentration of 0.3 U/μl]

2.5 µl of primer (30 µM or 300 µM to give a final concentration
of 3–30 µM)

2.5 µl of template (to give a final concentration of 0.1–1 ng/µl)

(Shorter primers require higher template concentrations and simple
(low complexity) genomes require higher Mg concentrations.)

- Centrifuge briefly to mix contents of tube.
- Add mineral oil (if necessary for thermal cycler).
- Cycle at 96°C for 20 s and 30°C for 20 s; 35 times.
- Dilute the samples 5- to 10-fold, and analyse by electrophoresis on
  polyacrylamide gels (described below) with detection by silver
  staining (described below).

5.4.3 SPECIFIC PCR

PCR of specific targets requires the design of appropriate primers and
the selection of appropriate amplification conditions. Primer design
can be assisted by the use of software to evaluate possible primers for
melting/annealing temperature and for any possible unwanted priming
sites of interactions. The annealing temperature and number of cycles
of amplification will often require optimization. The annealing temperature can be slowly increased until amplification fails to establish
practical annealing temperatures for reliable and specific amplification.
Some templates with a high GC content are more efficiently amplified
by including glycerol in the reaction (Henry and Oono, 1991).

(a) Example protocols

Species-specific PCR method

The following procedure was developed for the identification of species-
specific primers from the 5S RNA genes of plants (Ko and Henry, 1996).

- Amplify the 5S RNA spaces using the primers 5'-TTT AGT GCT
  GGT ATG ATC GC-3' and 5'-TGG GAA GTC CTC GTG TTG
  CA-3'.
- Excise the smallest or most distinct band from a 1.5% low melting
  agarose gel using phenol–chloroform and purifying using a micro-
  concentrator.
- Sequence directly or, if this proves unsuccessful, clone using a TA
  cloning method and sequence with M13 universal and reverse
  primers.
- Design a primer from the central part of the sequence and compare
  with sequences from related species to ensure specificity.

- Amplify a species-specific band using the primer with one of the original 'universal' primers used to amplify the spacer.

Screening transgenic plants for the presence of the bar gene using PCR

The *bar* gene from *Streptomyces hygroscopicusis* encodes phosphino-thricin acetyl transferase (PAT). The introduction of this gene during plant transformation provides the plant cell with resistance to the herbicide, bialaphos (see Chapter 4). Confirmation of transformation of herbicide-resistant plants or tissues can be provided by screening for the presence of the gene by PCR. PCR detection of this genes is made difficult by a high GC content and the presence of blocks of continuous GC sequences. The following procedure (developed by J.E. Vickers and G.C. Graham, unpublished results) has been found to be reliable.

- Combine 1 µg of genomic DNA, 2 µM of each primer (5'-CAGGAACCGCAGGAGTGGA-3' and 5'-CCAGAAACCCACGT-CATGCC-3', 200 µM dATP, dCTP, dGTP, dTTP, 1.5 mM $MgCl_2$, 2.6 U of Expand™ High Fidelity DNA polymerase and buffer in 25 µl.
- Denature at 94°C for 3 min and cycle 35 times at 55°C for 2 min, 72°C for 2 min and 94°C for 1 min, ending with 5 min at 72°C.
- Analyse PCR products for electrophoresis on a 1.5% agarose gel. Ethidium bromide staining should reveal a 372-base-pair product in plants that have been transformed with the *bar* gene.

5.5 Gel electrophoresis

5.5.1 AGAROSE GEL ELECTROPHORESIS

Agarose gel electrophoresis is the simplest method for the general analysis of plant DNA. This technique reveals the size distribution of DNA in any sample together with the approximate concentrations of the DNA components in the sample. The concentration of agarose in the gel can be varied to separate DNA molecules of different sizes (Table 5.1). A wide variety of agarose gel sizes can be used. The gel length, width and thickness can be varied with different commercial apparatus and the sample well spacings and sizes varied using different combs.

Electrophoresis technique

- Heat agarose in electrophoresis buffer (TAE, 0.04 M Tris–acetate, 0.001 M EDTA, pH 8.0) using a microwave oven until the agarose dissolves.

Table 5.1 Separation of DNA in agarose gels

Agarose concentration (%w/v)	Optimal separation range (kb)
0.3	5–60
0.6	1–20
0.7	0.8–10
0.9	0.5–7
1.2	0.4–6
1.5	0.2–3
2.0	0.1–2

- Add ethidium bromide (stock solution 10 mg/ml in water) to give 0.5 μg/ml and mix.
- Pour into gel tray and add comb to form sample wells.
- After the gel has set, remove the comb, and place gel into apparatus; cover gel with electrophoresis buffer.
- Load samples in gel-loading buffer [e.g. 0.25% bromophenol blue, 0.25% xylene cyanol FF, 40% (w/v) sucrose].
- Apply voltage (1–5 V/cm) to gel and leave until the dyes have migrated to the required extent.
- Examine gel under UV light and photograph to record results.

(b) Staining gels with ethidium bromide

Ethidium bromide can be used to stain nucleic acids in agarose gels; the dye allows visualization of DNA and RNA in gels under UV light. **Caution**: ethidium bromide is handled as a mutagen and toxin using gloves (Sambrook *et al.*, 1989). The procedure is as follows:

- Prepare a stock solution of ethidium bromide in water (10 mg/l). Ethidium bromide solutions should be stored in a dark place (e.g. the bottle should be covered with aluminium foil) at room temperature.
- Incorporate into the gel and electrophoresis buffer at 0.5 μg/ml.
- Examine gels using UV light. **Caution**: eye protection must be used and exposure of skin to UV light should also be avoided.

As an alternative, the gel may also be stained after electrophoresis by soaking briefly in a solution of ethidium bromide (0.5 μg/ml) in electrophoresis buffer.

5.5.2 POLYACRYLAMIDE GEL ELECTROPHORESIS

Polyacrylamide gel electrophoresis provides greater resolution than agarose gel electrophoresis in the analysis of small DNA fragments but is generally more difficult to perform.

(a) Electrophoresis technique

The following procedure is based upon the protocol of Caetano-Anolles and Bassam (1993) and is designed for use with the Mini-Protean II cell (Bio-Rad).

- Assemble electrophoresis apparatus, with a polyester backing film with the hydrophobic side up and place in casting stand.
- Prepare solutions for two 5% polyacrylamide gels by adding 4.2 g of urea, 1 ml of 10 × TBE (Tris–borate–EDTA) buffer (1 M Tris–HCl, 0.83 M boric acid, 10 mM EDTA, pH 8.3) and 1.2 ml of acrylamide stock (38% acrylamide, 2% piperazine diacrylamide) and make up to 10 ml with distilled water. **Caution:** acrylamide should be handled carefully as it is a neurotoxin.
- Add 150 μl of 10% ammonium persulphate (prepared freshly each day) and 15 μl of TEMED (*N,N,N',N'*-tetramethylethylenediamine) to initiate gel polymerization.
- Immediately fill gel using a syringe and place comb in position. The gel should be fully polymerized in 30 min.
- Assemble apparatus and fill buffer tanks with 1 × TBE.
- Remove comb and rinse wells with running buffer.
- Load samples (diluted in loading solution, 5 M urea, 0.02% xylene cyanol FF to contain 30–40 ng DNA in the 3 μl loaded) and electrophorese at 100 V for 60–90 min or until the dye front reaches three-quarters of the gel length.
- Remove gels for staining.

(b) Staining gels with silver

The following procedure for silver staining of polyacrylamide gels is based upon the procedure of Bassam and Caetano-Anolles (1993).

- Fix gels by immersing in 7.5% (v/v) acetic acid for 10 min.
- Wash gels three times in deionized water, each for 2 min.
- Stain with silver solution for 20 min. The solution should be prepared freshly by adding 1 g/l silver nitrate and 1.5 ml/l formaldehyde to water and stirring. **Caution:** this solution is toxic and should be handled and disposed of carefully.
- Rinse gels twice with distilled water.
- Develop until image is optimal (about 4 min). The developing

solution should be prepared freshly by adding 30 g of sodium carbonate, 3 ml of formaldehyde and 2 g of sodium thiosulphate to
water and mix by stirring. The solution should be placed on ice for
about 15 min before use and should be at about 8–12°C when used.

- Stop by washing in 7.5% (v/v) acetic acid.
- Wash with water.
- Gels may be preserved to produce a permanent record by drying
  at room temperature.

5.5.3 PULSED FIELD ELECTROPHORESIS

Pulsed field electrophoresis is used to separate and analyse large fragments of DNA. The method is used to fractionate DNA for cloning
in bacterial artificial chromosome (BAC) vectors.

5.6.1 DNA SEQUENCING

5.6 DNA sequencing and sequence analysis

DNA sequencing is an essential tool in the application of plant molecular biology. Sequencing is becoming an increasingly automated
process because of the large amount of data generated and the routine
nature of the need to obtain sequence information. Initial procedures
involved the automation of the manual method of dideoxy sequencing
(Figure 5.4). Further developments may be based on totally new
approaches to sequencing.

(a) DNA sequencing technique

The following procedure for cycle sequencing uses the ABI PRISM™
Dye Terminator.

- Prepare template DNA; plasmids may be isolated by alkaline
  analysis and PEG purification. PCR products may be purified using
  Centricon-100 columns.
- Combine 0.25–0.5 µg double-standard DNA template, 3.2 pmol of
  a sequencing primer, and 8 µl of a Terminator mix containing A-
  Dye Terminator, C-Dye Terminator, G-Dye Terminator, T-Dye
  Terminator, dITP, dATP, dCTP, dTTP in Tris–HCl, pH 9.0, $MgCl_2$,
  thermal stable pyrophosphatase and *Taq* DNA polymerase, in a
  total volume of 20 Ul.
- Cycle 25 times at 96°C for 10 s, 50°C for 5 s, and 60°C for 4 min
  in a Perkin-Elmer 9600.
- Transfer to a standard 1.5-ml microcentrifuge tube containing 2 µl
  of 3 M sodium acetate, pH 4.6 and 50 µl of 95% ethanol.

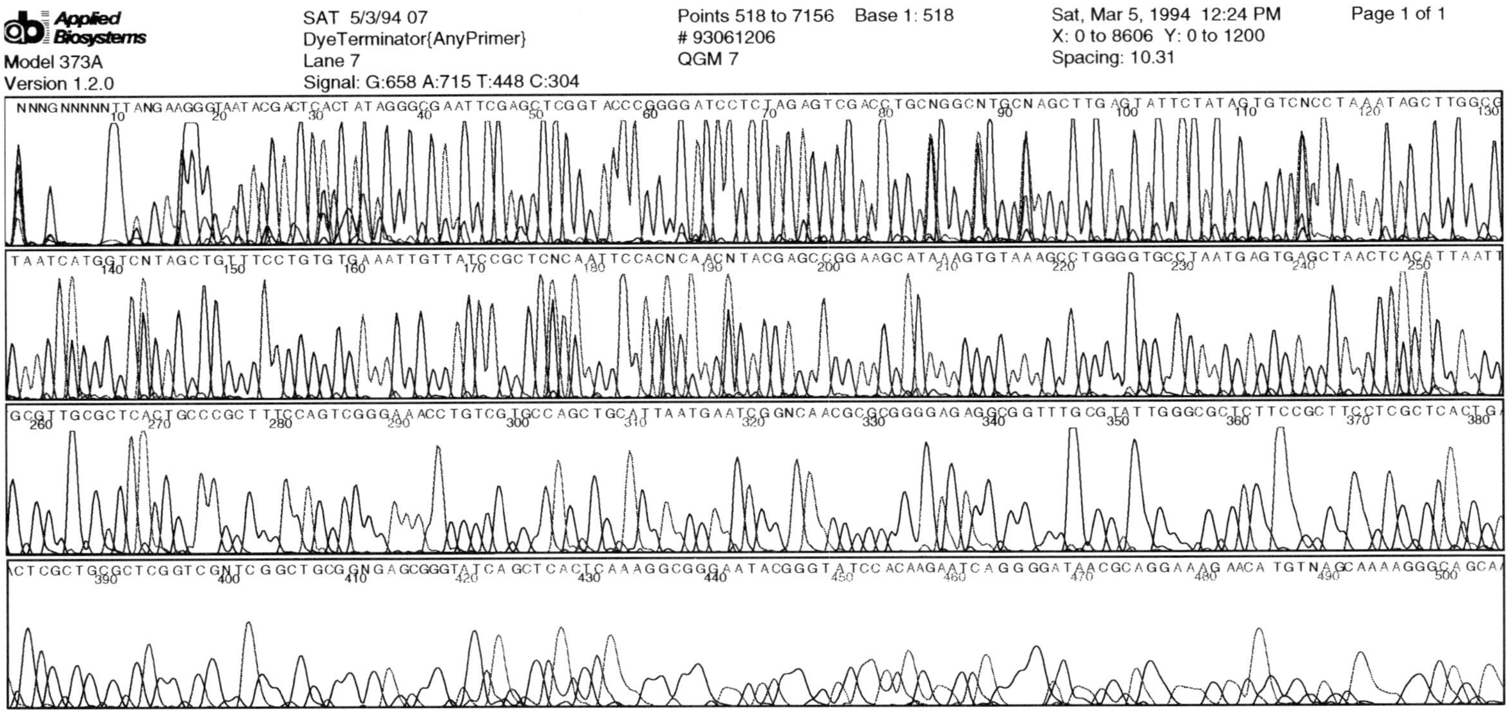

Figure 5.4 DNA sequencing. Output from automated DNA sequencing.

- Mix (vortex) and place on ice for 10 min.
- Centrifuge for 15–30 min at maximum speed available in micro-centrifuge or $12\,000 \times g$.
- Remove supernatant by aspiration.
- Rinse with 250 µl of 70% ethanol.
- Remove supernatant by aspiration and wipe sides of tube.
- Dry the pellet under a vacuum.
- The sample is now ready for loading after adding $5:1$ formamide/25 mM EDTA (pH 8.0) : 50 mg/ml Blue Dextran and heating at 90°C for 2 min.

5.6.2 SEQUENCE ANALYSIS

A wide range of software is available for the analysis of sequence data. Standard analyses include a search for open reading frames and the prediction of the amino acid sequence of proteins that could be encoded by the sequence (see Box). Comparisons of sequences of newly cloned genes with databases may allow a possible function to be attributed to the gene or the identity of the gene to be confirmed (Gaëta, 1995). However, the identification of gene function from sequence information alone remains difficult (Oliver, 1996).

5.7.1 INTRODUCTION

Cloning in plasmid vectors (Figure 5.5) has a wide range of practical applications in plant molecular biology. Plasmids are small double-stranded circular DNA molecules that are easily manipulated and propagated in bacteria. This makes them ideal vectors for work with recombinant DNA (Figure 5.6).

5.7 Cloning in plasmid vectors

(a) Plasmid preparation

Small-scale preparations of plasmid can be prepared by alkaline lysis.

- Grow a single colony of bacteria in a few ml of LB medium (10 g of bacto-tryptone, 5 g of bacto-yeast extract and 10 g of NaCl in 1 l of water adjusted to pH 7.0 with NaOH) containing an appropriate antibiotic, overnight at 37°C with vigorous shaking. A large vessel should be used (e.g. a 15-ml tube) with a loose cover.
- Collect the bacteria in a microfuge tube by centrifugation at $12\,000\,g$ for 30 s. Retain a small amount of the culture at 4°C. The following steps should be performed with the tubes stored on ice.

Sequence analysis

The example given is the genomic sequence of the bifunctional α-amylase/subtilisin inhibitor (BASI) from barley (Henry *et al.,* 1992). The sequence given in (a) was amplified by PCR using the primers underlined. The primers were designed from the cDNA sequence. The analysis indicates an absence of introns. The predicted amino acid sequence can be analysed as in the hydrophilicity plot shown in (b) for the predicted signal peptide.

(a)

```
                                             CTC   3
GAGGACACTCCAGCAGAGGTTTCAGTCATGGGTAGCCGCCGTGCAGGCCTCCT
                             MetGlySerArgArgAlaGlyLeuLeu
CTCCTCTCCCTTATTCTGGCCAGCACCGCCCTCTCGCGCAGC·GCCGATCCGCCG   111
LeuLeuSerLeuIleLeuAlaSerThrAlaLeuSerArgSer·AlaAspProPro
CCGGTGCACGACACGGACGGCCACGAGCTGCGCGCCGACGCCAACTACTACGTC
ProValHisAspThrAspGlyHisGluLeuArgAlaAspAlaAsnTyrTyrVal
CTCTCGGCCAACCGCGCCCACGGCGGGGGGGCTGACGATGGCGCCCGGCCACGGG   219
LeuSerAlaAsnArgAlaHisGlyGlyGlyLeuThrMetAlaProGlyHisGly
CGCCACTGCCCGCTCTTCGTGTCGCAGGACCCCAACGGGCAGCACGACGGGTTC
ArgHisCysProLeuPheValSerGlnAspProAsnGlyGlnHisAspGlyPhe
CCCGTGCGCATCACCCCGTACGGCGTCGCGCCGTCCGACAAGATCATCCGGCTG   327
ProValArgIleThrProTyrGlyValAlaProSerAspLysIleIleArgLeu
TCGACCGACGTCCGCATCTCCTTCCGCGCCTACACGACGTGTCTGCAGTCCACT
SerThrAspValArgIleSerPheArgAlaTyrThrThrCysLeuGlnSerThr
GAGTGGCACATCGACAGCGAGCTGGCGGCGGGCCGCCGGCACGTGATCACCGGC   435
GluTrpHisIleAspSerGluLeuAlaAlaGlyArgArgHisValIleThrGly
CCGGTCAAGGACCCGAGCCCGAGCGGCAGGGAGAACGCCTTCCGCATCGAGAAG
ProValLysAspProSerProSerGlyArgGluAsnAlaPheArgIleGluLys
TACAGCGGCGCCGAGGTGCACGAGTACAAGCTGATGTCGTGCGGGGACTGGTGC   543
TyrSerGlyAlaGluValHisGluTyrLysLeuMetSerCysGlyAspTrpCys
CAGGACCTCGGCGTGTTCAGGGACCTCAAGGGTGGGGCGTGGTTCTTGGGCGCC
GlnAspLeuGlyValPheArgAspLeuLysGlyGlyAlaTrpPheLeuGlyAla
ACCGAGCCATACCATGTCGTCGTGTTCAAGAAGGCGCCGCCCGCTTAAGGTCCA   651
ThrGluProTyrHisValValValPheLysLysAlaProProAla
ATGATCCATCCGTCAAGCGTGCGC
```

(b)

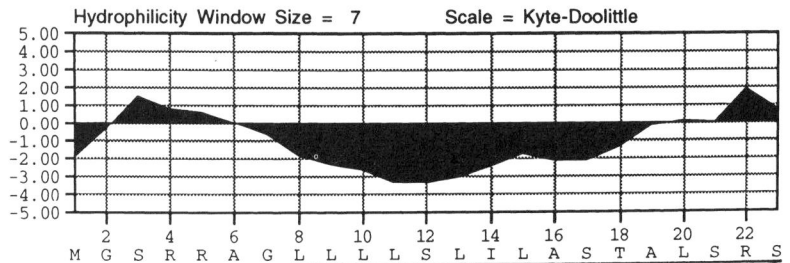

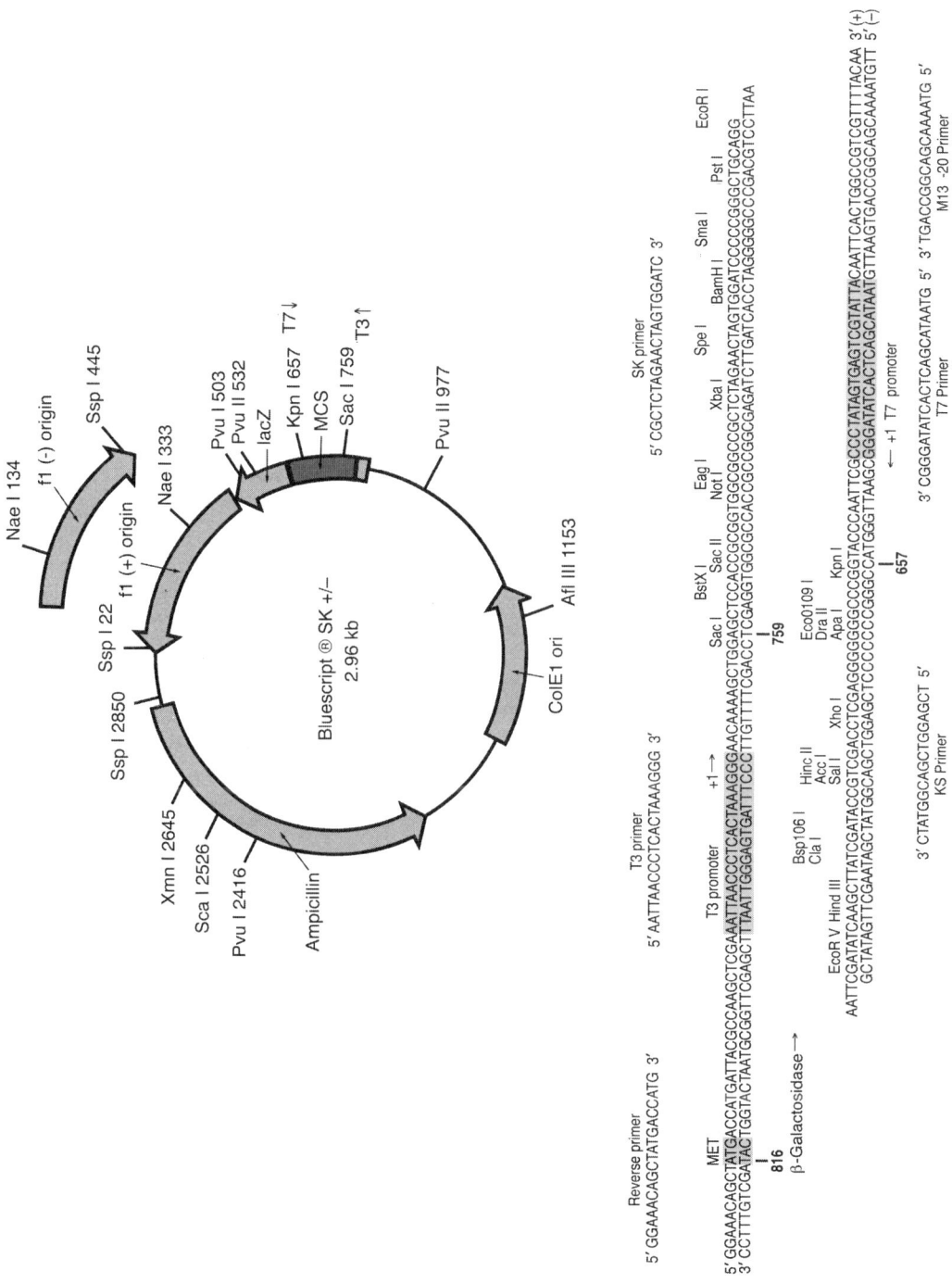

Figure 5.5 Bluescript. A plasmid vector for general use. The sites at which the plasmid is cut by restriction enzymes are marked. The poly or multiple cloning site (MCS) from *Sac*I to *Kpn*I allows insertion of the fragment to be cloned.

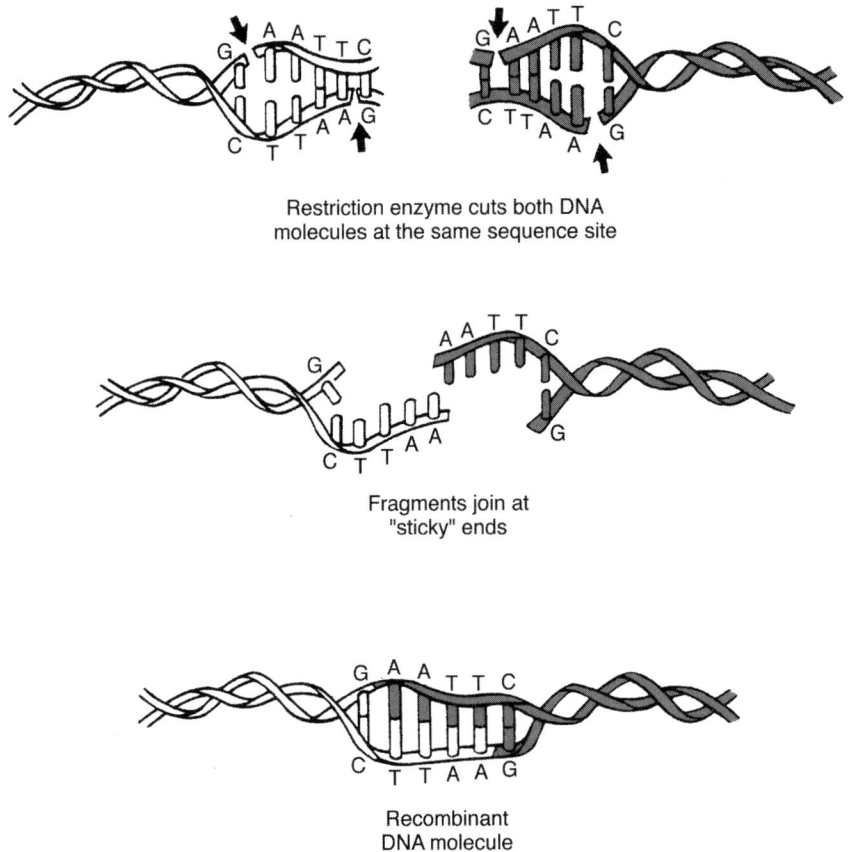

Restriction enzyme cuts both DNA
molecules at the same sequence site

Fragments join at
"sticky" ends

Recombinant
DNA molecule

Figure 5.6 Recombinant DNA techniques involve the cutting and joining of DNA molecules.

- Remove the supernatant with a pipette.
- Resuspend the bacteria in 100 μl of an ice-cold solution containing 50 mM glucose, 25 mM Tris–HCl, pH 8.0 and 10 mM EDTA. (This solution can be prepared in advance, autoclaved and stored at 4°C). Vigorous mixing using a vortex mixer may be necessary to completely resuspend the pellet.
- Add 200 μl of a freshly prepared solution of 0.2 N NaOH, 1% SDS (prepared from stable stock solutions). Close the cap of the tube and invert the tube five times, making sure that the entire surface of the tube is exposed to the solution.
- Add 150 μl of an ice-cold solution made by mixing 60 ml of 5 M potassium acetate, 11.5 ml of glacial acetic acid, and 28.5 ml of water. Close the tube and mix in an inverted position for 10 s.

Store on ice for 3–5 min.
- Centrifuge at 12 000 g for 5 min at 4°C and carefully transfer the supernatant to a fresh tube.
- Extract with phenol : chloroform by adding an equal volume, centrifuging at 12 000 g and 4°C for 2 min and transferring the upper phase to a fresh tube.
- Precipitate by adding 2 volumes of ethanol, vortexing and allowing to stand at room temperature for 3 min.
- Collect by centrifugation at 12 000 g for 5 min at 4°C.
- Remove the supernatant with a pipette and allow to drain upside down on a paper towel.
- Wash pellet with 1 ml of 70% ethanol at 4°C. Remove the supernatant as described above.
- Add TE buffer (10 mM Tris–HCl, pH 8.0, 1 mM EDTA) (50 μl) and store at –20°C. The volume may be adjusted depending on subsequent use. Vortex to aid dissolution of the pellet. RNase (free from DNase) can be added to remove contaminating RNA.
- Add two volumes of ethanol, vortex and allow to stand for 2 min at room temperature before centrifuging at 12 000 g for 5 min.
- Remove the supernatant with a pipette and drain by inverting the tube on a paper towel. Ensure that all drops of the supernatant are removed from the walls of the tube.
- Rinse the pellet with 70% ethanol and remove the supernatant as above.
- Add 50 μl of TE buffer (10 mM Tris–HCl, pH 8.0, 1 mM EDTA) and store at –20°C.

The volume can be adjusted depending on the requirements of subsequent protocols. RNase (free from DNase) can be added to remove contaminating RNA.

(b) Dephosphorylation

Dephosphorylation of the plasmid vector with alkaline phosphatase prevents ligation of the cut ends of the plasmid vector without the presence of an insert.

- Digest 10–20 μg of plasmid with desired restriction enzyme and resuspend digested plasmid in 90 μl of 10 mM Tris–HCl, pH 8.3.
- Add:
    10 μl of 10 × calf intestinal alkaline phosphatase buffer (10 mM $ZnCl_2$, 10 mM $MgCl_2$, 100 mM Tris–HCl, pH 8.3)
    1 unit of calf intestinal alkaline phosphatase (greater quantities may be needed for blunt or recessed termini)
- Incubate at 37°C for 1 h (for blunt or recessed termini the last 45 min should be at 55°C).

- Add SDS to 0.5%, EDTA (pH 8.3) to 5 mM and proteinase K to 100 μg/ml and incubate at 56°C for 30 min. This is designed to destroy the phosphatase completely to prevent interference in ligation reactions.
- Cool to room temperature and extract with phenol and then phenol : chloroform.
- Add 0.1 volume of 3 M sodium acetate (pH 7.0) and precipitate with 2 volumes of ethanol; wash with 70% ethanol and resuspend in TE buffer.
- Store aliquots at −20°C.

(c) Ligation

Cloning in plasmid vectors may be blunt-ended or involve the ligation of cohesive termini. Ligation of the vector and insert is achieved by using DNA ligase. Two separate control reactions should be performed with the vector alone and the foreign DNA alone.

- Mix 0.1 μg of the cut plasmid vector and an equimolar amount of the foreign DNA to be inserted in a total volume of 7.5 μl of water in a microfuge tube.
- Heat at 45°C for 5 min (to melt any cohesive termini that have reannealed) and chill on ice.
- Add:
  1 μl of 10 × bacteriophage T4 ligase buffer
  1 unit of DNA ligase
  1 μl of 5 mM ATP
- Incubate at 16°C for 12–16 hours (overnight).
- Transform competent *E. coli* with 1–2 μl of the ligation mixture.

(d) Transformation of competent *E. coli*

Preparation of competent *E. coli* may be achieved by treatment with calcium chloride. Competent *E. coli* may be transformed by heat shock.

- Remove a tube of frozen, competent cells from the −70°C freezer, thaw, and store on ice for 10 min.
- Add the DNA to the competent cells on ice for 30 min and swirl the tubes gently to mix the contents. The volume of DNA added should not exceed 5% of the volume of the cells (e.g. no more than 2.5 μl per 50 μl of competent cells) and contain approximately 1 ng of supercoiled plasmid DNA.
- Incubate at 42°C in a water bath for exactly 90 s, without shaking.
- Place the tubes on ice for 1–2 min to chill.
- Add 0.8 ml of growth medium (SOC) and incubate at 37°C for 45 min (preferably with gentle shaking) to allow recovery and expression of antibiotic resistance.

- Plate on appropriate media containing antibiotic and incubate at 37°C for 12–16 hours.

(e) Blue–white screening

Vectors with a polycloning site within the coding region of a part of a β-galactosidase gene (*lacZ*) allow blue–white screening. The 146 N-terminal amino acids encoded can associate with the carboxy-terminal region of β-galactosidase produced by appropriate E. coli hosts to produce an active β-galactosidase. Insertion of a DNA fragment in the polycloning site disrupts the production of an effective peptide, preventing complementation to form an active β-galactosidase. Vectors containing an insert can thus be detected by the loss of β-galactosidase activity:

- Add 40 µl of a stock of X-gal* (5-bromo-4-chloro-3-indolyl-β-D-galactosidase) and 4 µl of a stock of IPTG* (isopropylthio-β-D-galactosidase) to an LB plate.
- Spread with a glass spreader and allow to dry.
- Inoculate with the bacteria to be screened and incubate at 37°C for 12–16 hours.
- Colonies containing plasmid with no insert should appear blue, and those with an insert should be white. Pick white colonies using a toothpick and test by preparing the plasmid by alkaline-lysis. Colonies may also be screened by PCR.

5.7.2 CLONING OF PCR PRODUCTS

Cloning of PCR products is a common requirement in the production of gene constructs, useful probes and the sequencing of some PCR products. PCR products generated with a single primer, such as in RAPD analysis, cannot be sequenced directly because of the presence of the same sequence at both ends and the lack of other sequence information on which to base the design of sequencing primers. Cloning of these PCR products is necessary for their sequencing.

Blunt end cloning has been widely used to clone PCR products. PCR products may be treated with Klenow DNA polymerase to ensure that all products are blunt ended before ligation. PCR with DNA polymerase from *Thermus aquaticus* (*Taq*), results in products with a 3'-A overhang because of the enzyme's ability to add a single A to a double-stranded DNA. Vectors with a complementary T can be used to clone

---

*Stock solutions of X-gal (20 mg/ml in dimethyl formamide) should be stored wrapped in aluminium foil at –20°C. IPTG stock solutions (2 g IPTG in 8 ml of water) should be stored frozen at –20°C.

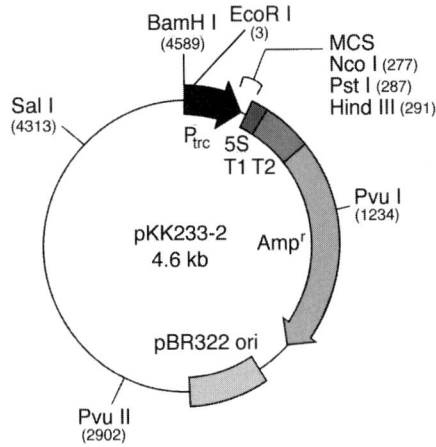

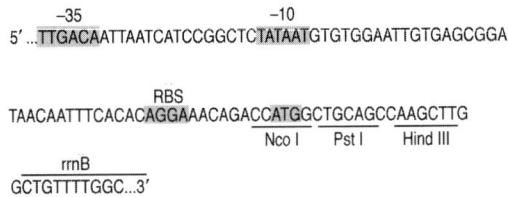

Figure 5.7 pKK233 vector for expression of proteins. This vector contains the *trp-lac* (*tac*) promoter that may be derepressed by the addition of isopropyl-β-D-galactoside (IPTG).

these products with high efficiency. This approach is known as TA cloning (Zhou *et al.*, 1995).

5.7.3 EXPRESSION OF PLANT GENES IN BACTERIA

Cloning in expression vectors may used to generate large amounts of plant proteins in bacteria. These proteins may be useful in food or industrial processes or may be used in characterizing the protein. This approach may be of special value when the protein is not abundant in the plant or is difficult to purify. Vectors are available for direct expression of the protein (Figure 5.7); however, expression as a fusion protein (Figure 5.8) may have advantages in purification of the protein or when the native protein is toxic to the bacteria.

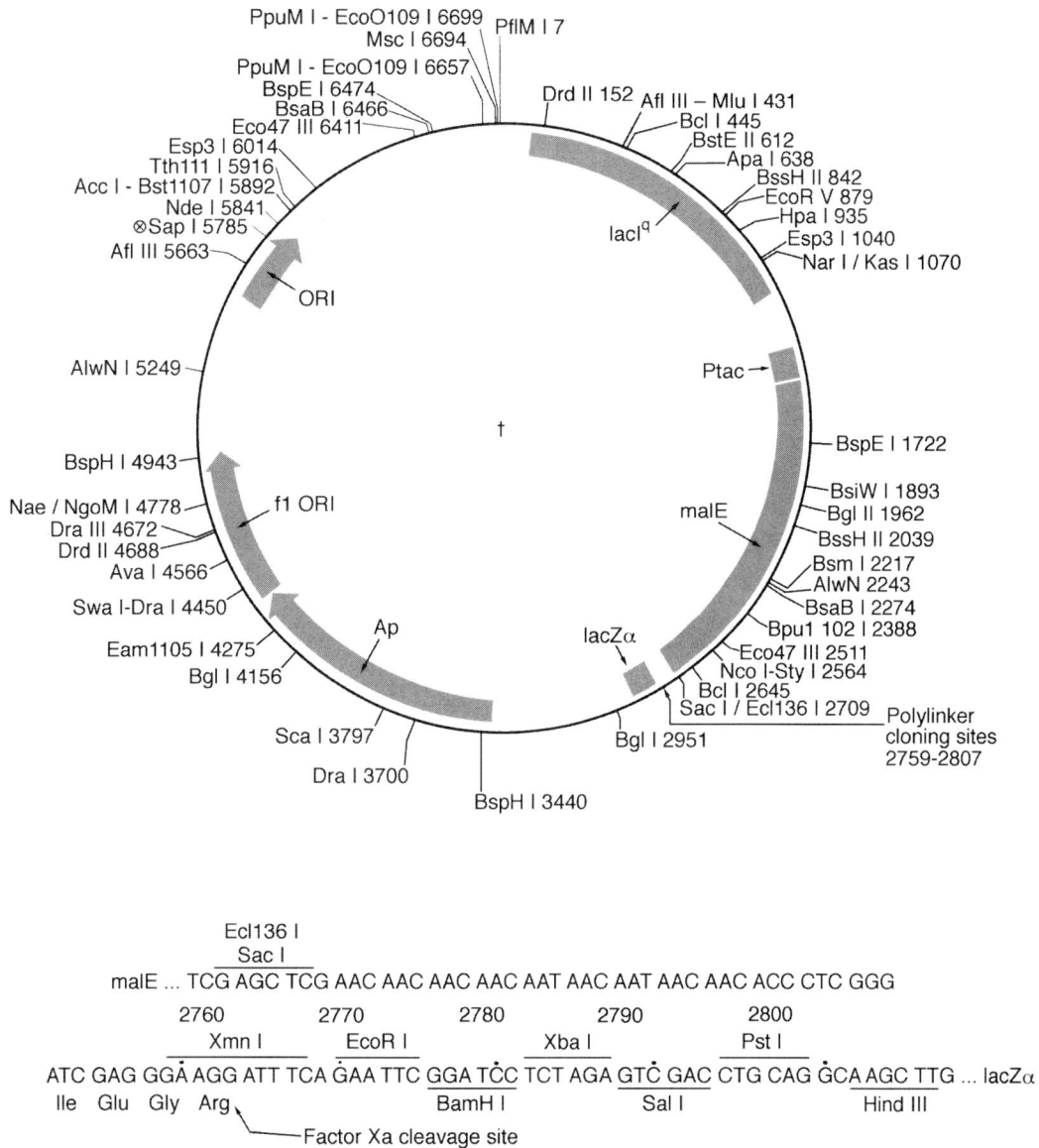

Figure 5.8 pMAL vector for expression and purification of proteins. This vector contains the *tac* promoter and is designed to express a maltose-binding protein–fusion protein. (Copyright © 1996/97. New England Biolabs catalog. Reprinted with pemission)

Bacterial expression of a plant protein

Bacterial expression of bifunctional α-amylase/subtilisin inhibitor (BASI) from barley using the pMAL system (Jones *et al.*, 1995).

(a) SDS–PAGE analysis of induction and expression of fusion protein. The bacterial culture was grown at 37°C with vigorous shaking for 2 hours before being induced with IPTG. The cells were then harvested, freeze–thawed, and sonicated to release the protein. Lanes 1 and 8, molecular weight standards; lane 2; uninduced cells; lane 3; induced cells 1 h after induction; lane 4, induced cells at 2 h; lane 5, induced cells at 3 h; lane 6, crude extract containing soluble protein; lane 7, insoluble material. The arrow points to the fusion protein at 63 kDa.

(b) SDS–PAGE analysis of factor Xa cleavage. Lane 1, broad-range molecular weight standards; lane 2, uncut MBP-BASI fusion protein, lane 3, cleavage mix at 2 h; lane 4, cleavage mix at 4 h; lane 5, cleavage mix at 24 h; lane 6, cleavage mix at 3 days. Bands are of the expected size for both BASI and MBP (see arrows).

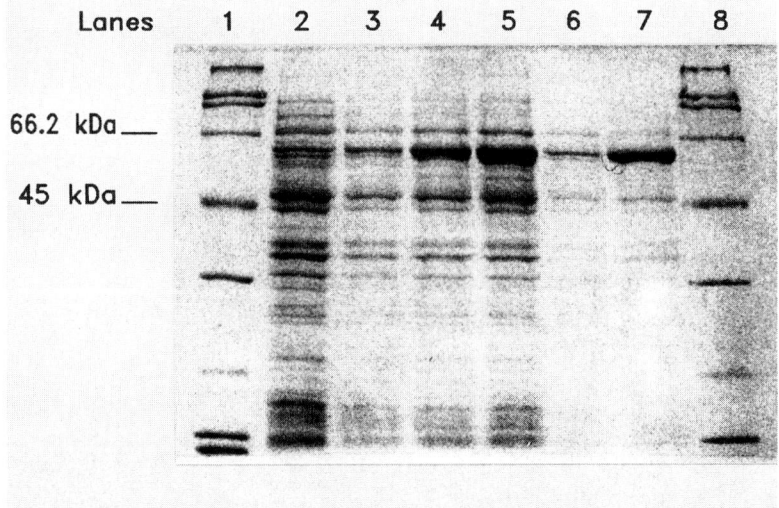

(a)

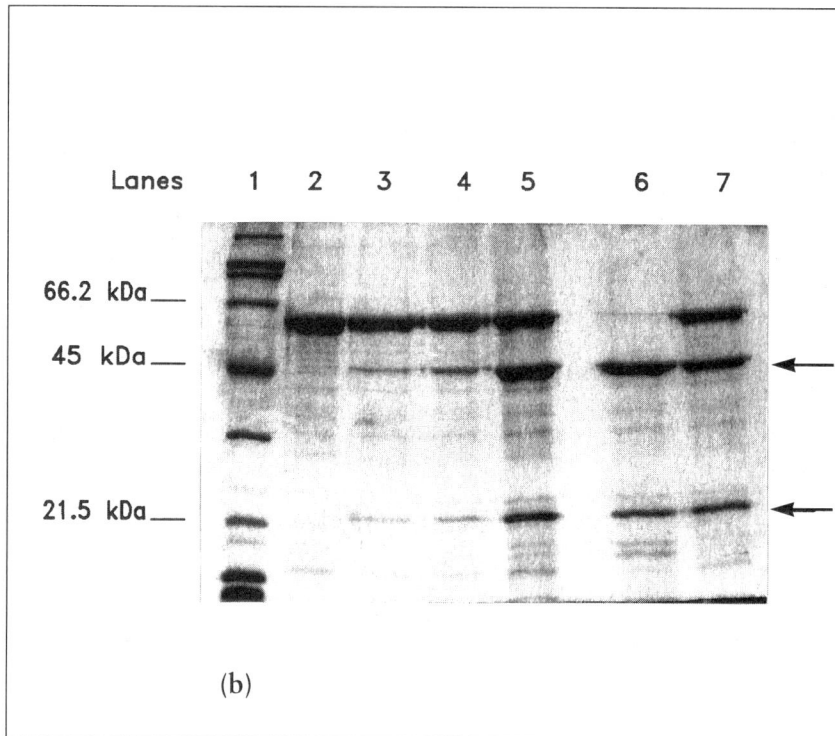

(b)

5.8.1 *AGROBACTERIUM*-MEDIATED

Transformation of plants with *Agrobacterium* may be highly efficient. Conditions for achieving high levels of transformation have been defined (Rempel and Nelson, 1995). Antibiotics most effective against *Agrobacterium* have been identified (Shackelford and Chlan, 1996).

5.8.2 MICROPARTICLE BOMBARDMENT

Microprojectile devices employing gunpowder have largely been replaced by gas-powered apparatus which allows better control of microparticle bombardment (Figures 5.9 and 5.10). The microparticles used are either tungsten or gold. The coating of the DNA on the microparticles is a crucial part of the procedure.

5.8 Plant transformation

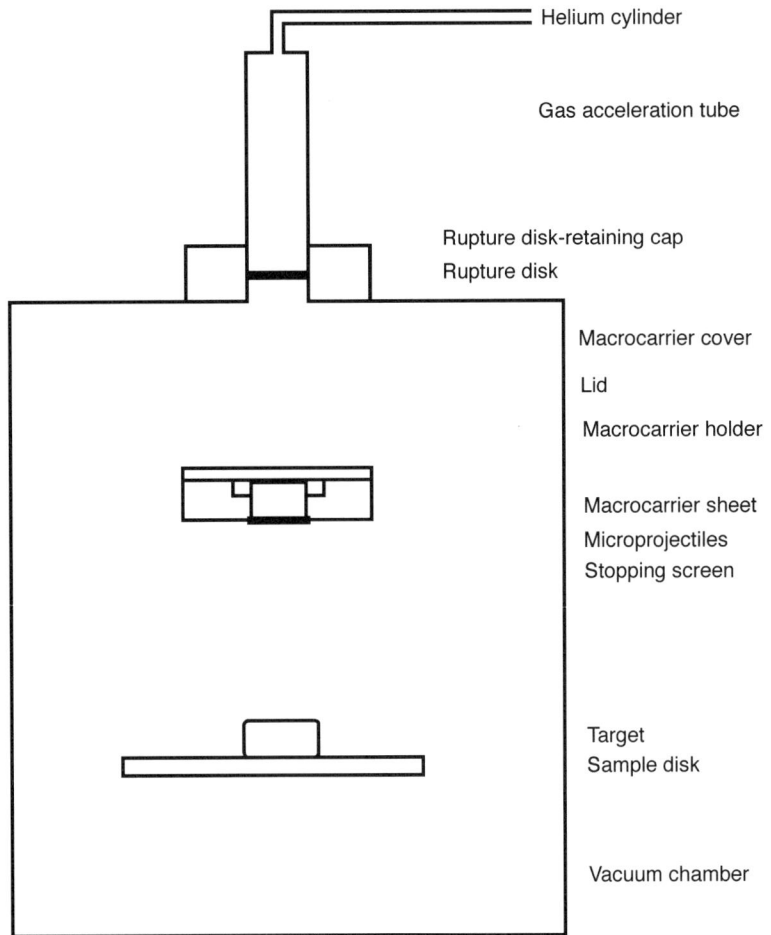

Figure 5.9 Schematic representation of a device for transformation of plants by microprojectile bombardment – the gas pulse gun.

(a) Microparticle bombardment technique

Optimization of bombardment conditions may include a consideration of the following:

- Choice of target tissue
- Physiological status of target tissue (e.g. osmotic stress)
- Size and type (gold or tungsten) of microparticle
- Concentration of DNA plasmid used to prepare coated particles
- Pressure or velocity used for microparticle projection
- Distance of target tissue from source of microparticles

Figure 5.10 Apparatus for microprojectile bombardment.

- Strength of vacuum used to evacuate chamber for bombardment
- Inclusion of selectable marker on reporter gene constructs
- Choice of regeneration and selection protocols

5.8.3 ASSAY OF MARKER GENES

The *uid* A gene from bacteria, encoding β-glucuronidase (GUS), has become the most widely used marker gene in plants. Commonly used constructs encoding this gene are shown in Figures 5.11 and 5.12. This marker is used in histochemical and quantitative assays with the colorimetric substrate, 5-bromo-4-chloro-3-indolyl-β-D-glucuronic acid (Figure 5.13) and with a fluorimetric substrate, 4-methylumbelliferyl-β-D-glucuronide. Low levels of endogenous activity in some plant tissues can be eliminated as a background by use of appropriate assays (Hansch *et al.*, 1995).

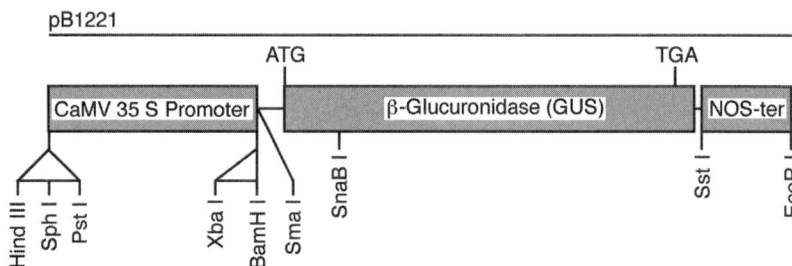

Figure 5.11 pBI221. A marker gene construct for use in plant transformation. GUS is expressed in a wide range of tissues with this construct.

(a) Assay of GUS (β-glucuronidase)

- Extract tissue by homogenization in 50 mM sodium phosphate buffer, pH 7.0, 10 mM EDTA, 0.1% Triton X-100, 0.1% sarkosyl, 10 mM β-mercaptoethanol. [Phenylmethyl sulphonyl fluoride (PMSF) can be added at 25 μg/ml if proteases are likely to be abundant.]
- High backgrounds may be reduced by passing the extract through a gel filtration column to remove low-molecular weight components that may interfere. (β-Glucuronidase from *E. coli* has a molecular weight of 74 000 Da.)
- Colorimetric assays may be performed by incubating the extract in 50 mM sodium phosphate buffer, pH 7.0, 10 mM β-mercapto-ethanol, 1 mM *p*-nitrophenyl glucuronide, 0.1% Triton X-100 at 37°C. A reaction of 1 ml volume can be stopped by adding 0.4 ml of 2-amino-2-methyl propanediol. The time of incubation can be extended to increase sensitivity, with overnight incubations giving good results. Absorbance at 415 nm gives a measure of GUS activity.
- Fluorimetric assays providing greater sensitivity may be conducted by substituting 1 mM 4-methyl umbelliferyl glucuronide for the colorimetric substrate. The other components of the reaction mixture are as for the colorimetric assays. A reaction in a total volume of 0.2 ml can be stopped by adding 1 ml of 0.2 M sodium carbonate. Fluorescence at 455 nm resulting from excitation at 365 nm is a measure of GUS activity.

Further details on GUS may be found in the GUS Gene Fusion System User's Manual (Jefferson, CLONTECH).

Firefly luciferase may be used as a marker gene. The activity is assayed by adding luciferin, Mg^{2+} and ATP to extracts or plant tissues. Light is produced by the luciferase and can be monitored using a lumi-nometer or a scintillation counter.

pBI121

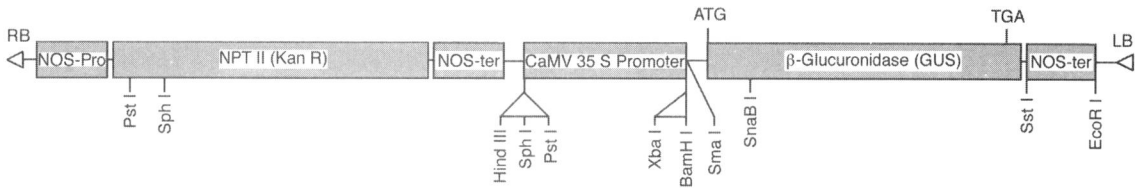

Figure 5.12 pBI121. Similar to pBI221, but including a gene for resistance to the antibiotic, kanamycin.

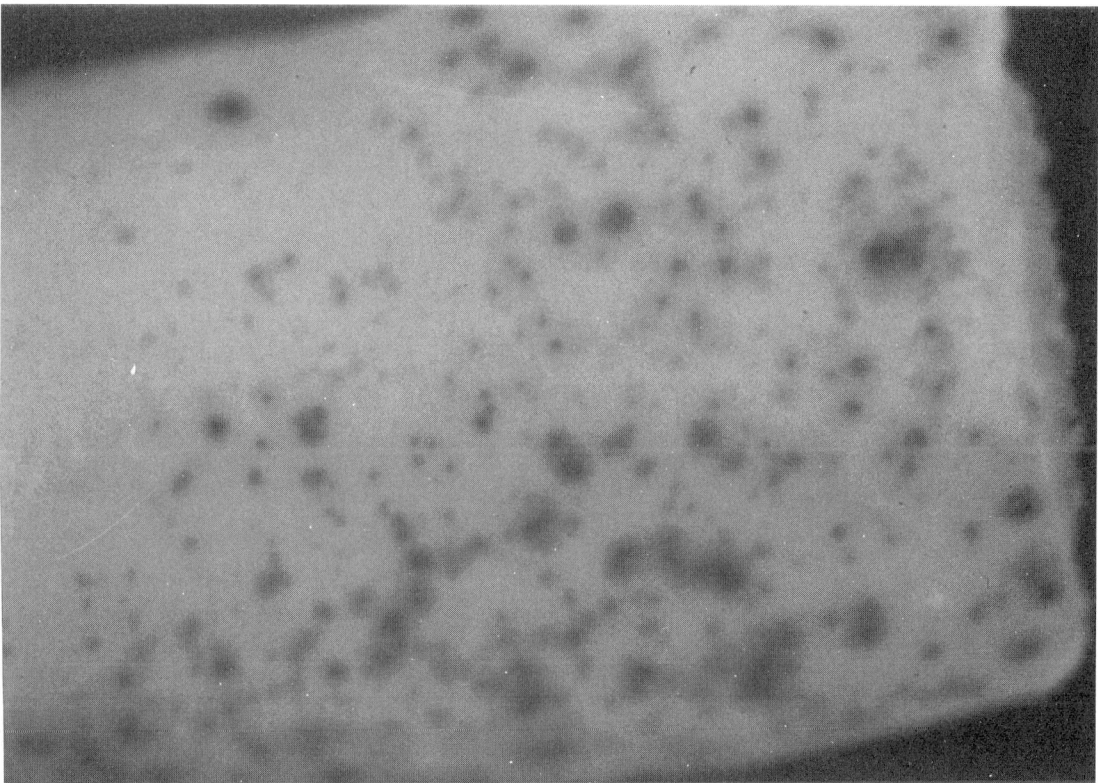

Figure 5.13 Expression of marker genes introduced by microparticle bombardment. Expression of GUS in cauliflower.

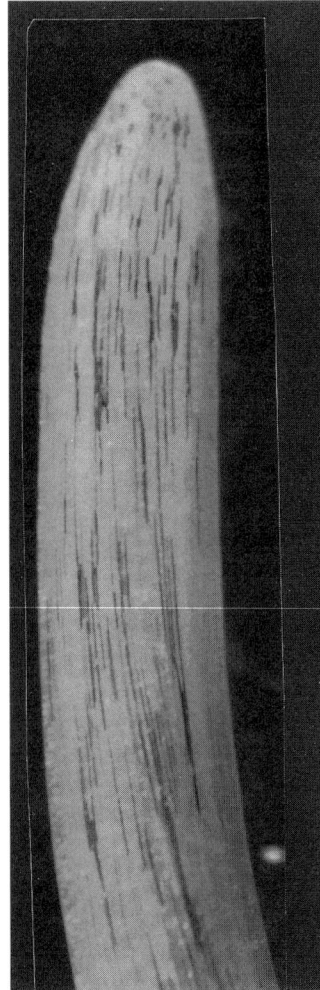

Figure 5.14 Expression of marker genes in coleoptile. The Cl and R genes were introduced into wheat by microparticle bombardment.

The green fluorescent protein (GFP) from the bioluminescent jelly-fish, *Aequorea victoria*, has been used as reporter gene in plants (Niedz *et al.*, 1995). Transformed cells emit an intense green light when illuminated in blue light (450–490 nm).

Some markers may be used only in specific species. For example, genes regulating anthocyanin synthesis in cereals have been used as visible marker genes. The C1 and R genes (transcriptional regulators) from maize directly cause the development of red coloration in tissues (Figure 5.14) without the need for any staining and have been used as marker genes in the cereals (Dhir *et al.*, 1994).

5.9.1 GENOMIC LIBRARY PREPARATION

The preparation of genomic libraries in bacteriophage lambda involves partial digestion of high-molecular weight DNA before cloning so as to generate a library that contains clones representing the genome as completely as possible. Genomic libraries for many species are available commercially and this may be an attractive alternative to generating your own.

5.9 Library production

5.9.2 CDNA LIBRARY PRODUCTION

The cloning of genes expressed in specific tissues or under particular developmental conditions has usually been achieved by the production of a cDNA library. Screening of the cDNA library allows the cloning of the cDNA of interest and this is often used in turn as a probe in screening to isolate a genomic clone from a genomic library. The procedure requires the isolation of polyadenylated RNA from the tissue of interest. A reverse transcriptase is used to produce a DNA copy of the RNA as a first strand. This is then converted to a double-stranded DNA in a second strand synthesis reaction involving RNase H and DNA polymerase I. The fragments are made blunt ended using Klenow enzyme and linkers are ligated to allow cloning into bacteriophage lambda. The isolation of undegraded mRNA is essential to the successful preparation of a cDNA library.

5.9.3 BAC AND YAC LIBRARY PRODUCTION

The maximum size of DNA fragments that can be cloned in bacterio-phage lambda is about 24 kb. The cloning of fragments larger than this requires the use of vectors such as cosmids (up to 45 kb) or BACs (bacterial artificial chromosomes) and YACs (yeast artificial chromo-somes). The first step in the preparation of BAC or YAC libraries is the isolation of very large DNA fragments (Woo *et al.*, 1995). The cell walls of plants make this a more difficult task than it is with mammalian tissues. Two approaches have been used: (i) isolation of nuclei; and (ii) isolation of protoplasts. The nuclei or protoplasts are then embedded in either agarose plugs or agarose microbeads. The embedded cells or organelles are then lysed and treated with protease to degrade protein. Partial digestion with *Hin*dIII is followed by size selection of fragments above 100 kb using pulsed field gel electro-phoresis for ligation into dephosphorylated BAC vectors. BAC clones with inserts may be identified by normal blue–white screening in vectors such as pBeloBAC11. BACs are maintained as single copy plasmids in

Figure 5.15 Screening of a genomic library in lambda using a cDNA probe.

E. coli. The main advantages of BACs relative to YACs are a lower level of chimerism and ease of preparation, maintenance and manipulation.

5.9.4 SCREENING OF LIBRARIES

cDNA and genomic libraries may be screened (Figure 5.15) by using DNA probes [synthetic oligonucleotides or cDNA (full-length or partial)], by PCR or by expression and the use of antibodies or other detection methods for the gene products. Protocols for screening with probes are related to those described in relation to Southern blotting with the full range of labelling and detection systems being available.

5.10 Cloning plant genes

A range of techniques are available for the cloning of plant genes. The approaches rely upon a knowledge of one or more of the following (Sawahel and Kiichi, 1995):

1. The sequence of the gene or gene products: screening of libraries with a probe (cDNA or synthetic oligonucleotide) allows the isolation of clones with homologous sequences.
2. The location of the gene in the genome: map-based cloning is possible when the location of the gene is known. This involves 'walking' from a known linked marker to the gene of interest.
3. The pattern of gene expression: differential screening of cDNA libraries may be used or subtraction of libraries may allow the isolation of cDNA clones unique to particular tissues of stages of development. Differential display (described below) has been used more recently for this purpose.
4. The function of the gene: genes may be cloned by analysis of their ability to add a function or trait to a line lacking the gene. Screening of genomic libraries in this way requires an efficient transformation system to generate the large numbers of transgenic lines required. Transposon insertion leading to a loss of function by gene inactivation has been more widely used for cloning on the basis of phenotype or known function.

Many other procedures may be of value in practical application of plant molecular biology. A few examples are given here with reference to sources of appropriate protocols.

> **5.11**
> **Miscellaneous**
> **methods**

(a) *In vitro* translation

The translation of plant mRNA *in vitro* may be of value in the identification or characterization of the mRNA. This technique may also be useful in studies of the mechanism and control of protein synthesis. Speirs (1993) gives protocols for *in vitro* translation systems.

(b) 5 '/3' RACE

Screening of cDNA libraries is rarely successful in recovering the full-length mRNA. Techniques are available for obtaining missing sequences from either the 5' or 3' end. These methods involve the rapid amplification of cDNA ends (RACE) using PCR with primers to internal known sequences and oligo-dT primers following tailing with dATP.

(c) Microsatellite markers

Microsatellite markers are identified by screening libraries that have been enriched in simple repeat sequences. PCR strategies have been devised (Thomas and Scott, 1994) for quickly obtaining the sequences of flanking regions for use in microsatellite analysis.

(d) Differential display

The technique of differential display may be used to isolate a specific mRNA (Liang and Pardee, 1992). PCR amplification with short primers of arbitrary sequence is used to identify differences in the mRNA from different tissues or at different developmental stages. The procedure involves first strand cDNA synthesis followed by PCR with an oligo-dT primer and a short primer of arbitrary sequence. Differentially displayed fragments can be cloned, sequenced and used as probes to isolate full-length cDNAs (Song *et al.*, 1995). Rare tissue or development-specific genes may be cloned in this way.

KEY TERMS

Plasmid
Southern blotting
Northern blotting
Western blotting
Restriction enzyme
Dideoxy sequencing
Ligation
Blue–white screening
cDNA
Bacterial Artificial Chromosome (BAC)
Yeast Artificial Chromosome (YAC)
Differential display

EXAMPLES OF WORKED QUESTION

1. How is plant material best collected for DNA extraction?

   Fresh or frozen material is best, but dry plant material may also be suitable. Frozen material should not be allowed to thaw before extraction.

2. What are the main alternatives for estimation of DNA?

   Absorbance in the UV, estimation from an ethidium bromide-stained gel or fluorimetric analysis.

3. What are Southern, Northern and Western blotting?

   Blotting of DNA, RNA and proteins respectively onto membranes for specific detection.

4. When would you use a YAC or a BAC vector?

   For the cloning of DNA fragments larger than about 4 kb (possible in cosmids).

5. List the types of information necessary to be able to clone a plant gene.

   Known sequence of gene or gene product
   or
   Known location of the gene in the genome
   or
   Known pattern of gene expression
   or
   Known function of gene

---

1. Outline the main considerations in selecting a DNA extraction procedure.

2. Describe the options available for the labelling of DNA.

3. How can the consistency of PCR analysis be improved?

4. Outline the protocols involved in cloning with plasmid vectors.

Questions

References

Apuya, N.R., Frazier, B.L., Keim, P., Jill Roth, E. and Lark, K.G. (1988) Restriction fragment length polymorphisms as genetic markers in soybean, *Glycine max* (L.) merill. *Theoretical and Applied Genetics*, **75**, 889–901.

Bassam, B.J. and Caetano-Anolles, G. (1993) Silver staining of DNA in polyacrylamide gels. *Applied Biochemistry and Biotechnology*, **42**, 181–8.

Berthomieu, P. and Meyer, C. (1991) Direct amplification of plant genomic DNA from leaves and root pieces using PCR. *Plant Molecular Biology*, **17**, 555–7.

Birch, D.E., Kolmodin, L., Laird, W.J., McKinney, N., Wong, J., Young, K.K.Y., Zangenberg, G.A. and Zoccoli, M.A. (1996) Simplified hot start PCR. *Nature*, **381**, 445–6.

Caetano-Anolles, G. and Bassam, B.J. (1993) DNA amplification fingerprinting using arbitrary oligonucleotide primers. *Applied Biochemistry and Biotechnology*, **42**, 189–200.

Chunwongse, J., Martin, G.B. and Tanksley, S.D. (1993) Pre-germination genotypic screening using PCR amplification of half-seeds. *Theoretical and Applied Genetics*, **86**, 694–8.

Clarke, B.C., Moran, L.B. and Appels, R. (1989) DNA analyses in wheat breeding. *Genome*, **32**, 334–9.

Dhir, S.K., Pajeau, M.E., Fromm, M.E. and Fry, J.E. (1994) Anthocyanin genes as visual markers for wheat transformation, in *Improvement of Cereal Quality by Genetic Engineering* (eds R.J. Henry and J.A. Ronalds), Plenum, New York, pp. 71–5.

Gaëta, B.A. (1995) Database similarity searching using BLAST and FastA *Australasian Biotechnology*, **5**, 282–90.

Graham, G.C., Mayers, P. and Henry, R.J. (1994) A simplified method for the preparation of fungal genomic DNA for PCR and RAPD analysis. *BioTechniques*, **16**, 48–50.

Hansch, R., Koprek, T., Mendel, R.R. and Schulze, J. (1995) An improved protocol for eliminating endogenous β-glucuronidase background in barley. *Plant Science*, **105**, 63–9.

Henry, R.J. and Oono, K. (1991) Amplification of a GC-rich sequence from barley by a two-step polymerase chain reaction in glycerol. *Plant Molecular Biology Reporter*, **9**, 139–44.

Henry, R.J., McKinnon, G.E., Haak, I.C. and Brennan, P.S. (1992) Use of alpha-amylase inhibitors to control sprouting, in *Preharvest Sprouting in Cereals 1992* (eds M.K. Walker-Simmons and J.L. Reid), American Association of Cereal Chemists, St. Paul, pp. 232–5.

Jarrett, S.J., Marschke, R.J., Symons, M.H., Gibson, C.E., Henry, R.J., and Fox G.P. (1996) Alpha-amylase/subtilisin inhibitor levels in Australian barleys. *Journal of Cereal Science* (in press).

Jones, M., Vickers, J., Henry, R.J., Symons, M., Marschke, R.J. and de Jersey, J. (1995) Bacterial expression of the bifunctional alpha-amylase/subtilisin inhibitor from barley. Cereals '95, 45th Australian Cereal Chemistry Conference Proceedings, Royal Australian Chemical Institute, Melbourne, pp. 449–52.

Ko, H.L. and Henry, R.J. (1996) Specific 5S ribosomal RNA primers for plant species identification in admixtures. *Plant Molecular Biology Reporter*, **14**, 33–43.

Langridge, U., Schwall, M. and Langridge, P. (1991) Squashes of plant tissue as a substrate for PCR. *Nucleic Acids Research*, **19**, 6954.

Liang, P. and Pardee, A.B. (1992) Differential display of eukaryotic RNA by means of the polymerase chain reaction. *Science*, **257**, 967–70.

Lin, J.-Z. and Ritland, K. (1995) Flower petals allow simpler and better isolation of DNA for plant RAPD analyses. *Plant Molecular Biology Reporter*, **13**, 210–13.

Marechal-Drouard, L. and Guillemaut, P. (1995) A powerful but simple technique to prepare polysaccharide-free DNA quickly and without phenol extraction. *Plant Molecular Biology Reporter*, **13**, 26–30.

Michaud, H., Lumaret, R., Ripoll, J.P., and Toumi, L. (1995) A procedure for the extraction of chloroplast DNA from broad leaved tree species. *Plant Molecular Biology Reporter*, **13**, 131–7.

Niedz, R.P., Sussman, M.R. and Satterlee, J.S. (1995) Green fluorescent protein: an *in vivo* reporter of plant gene expression. *Plant Cell Reporters*, **14**, 403–6.

Oliver, S.G. (1996) From DNA sequence to biological function. *Nature*, **379**, 597–600.

Pandey, R.N., Adams, R.P. and Flournoy, L.E. (1996) Inhibition of random amplified polymorphic DNAs (RAPDs) by plant polysaccharides. *Plant Molecular Biology Reporter*, **14**, 17–22.

Rempel, H.C. and Nelson, L.M. (1995) Analysis of conditions for *Agrobacterium*-mediated transformation of tobacco cells in suspension. *Transgenic Research*, **4**, 199–207.

Rether, B., Delmas, G. and Laouedj, A. (1993) Isolation of polysaccharide-free DNA from plants. *Plant Molecular Biology Reporter*, **11**, 333–7.

Sambrook, J., Fritsch, E.F. and Maniatis, T. (1989) *Molecular Cloning: A Laboratory Manual*, 2nd edition, Cold Spring Harbor Laboratory Press, Cold Spring Harbor, New York.

Sawahel, W. and Kiichi, F. (1995) Gene cloning in plants: innovative approaches. *BioTechniques*, **19**, 106–14.

Shackelford, N.J. and Chlan, C.A. (1996) Identification of antibiotics that are effective in eliminating *Agrobacterium tumetaciens*. *Plant Molecular Biology Reporter*, **14**, 50–7.

Song, P., Yamamoto, E. and Allen, R.D. (1995) Improved procedure for differential display of transcripts from cotton tissues. *Plant Molecular Biology Reporter*, **13**, 174–81.

Speirs, J. (1993) *In vitro* translation of plant messenger RNA. in Methods in Plant Biochemistry, Vol. 10, Academic Press, 33-56.

Steiner, J.J., Poklemba, C.J., Fjellstrom, R.G. and Elliott, L.F. (1995) A rapid one-tube genomic DNA extraction process for PCR and RAPD analyses. *Nucleic Acids Research*, **23**, 2569–70.

Thomas, M.R. and Scott, N.S. (1994) Sequence-tagged site markers for microsatellites: simplified technique for rapidly obtaining flanking sequences. *Plant Molecular Biology Reporter*, **12**, 58–64.

Thomson, D. and Henry, R. (1993) Use of DNA from dry leaves for PCR and RAPD analysis. *Plant Molecular Biology Reporter*, **11**, 202–6.

Thomson, D. and Henry, R. (1995) A rapid single-step protocol for preparation of plant tissue for analysis by PCR. *BioTechniques*, **19**, 394–400.

Williams, J.G.K., Kubelik, A.R., Livak, K.J., Rafalski, J.A. and Tingey, S.V. (1990) DNA polymorphisms amplified by arbitrary primers are useful as genetic markers. *Nucleic Acids Research*, **18**, 6531–5.

Woo, S-S., Rastogi, V.K., Zhang, H-B., Paterson, A.H., Schertz, K.F. and Wing, R.A. (1995) Isolation of megabase-size DNA from sorghum and application for physical mapping and bacterial and yeast artificial chromosome library construction. *Plant Molecular Biology Reporter*, **13**, 82–94.

Yu, K. and Pauls, K.P. (1994) Optimization of DNA-extraction and PCR procedures for random amplified polymorphic DNA (RAPD) analysis in plants, in *PCR Technology: Current Innovations* (eds H.G. Griffin and A.M. Griffin) CRC Press, Boca Raton, pp. 193–200.

Zhou, M-Y., Clark, S.E. and Gomez-Sanchez, C.E. (1995) Universal cloning method by TA strategy. *BioTechniques*, **19**, 34–5.

Appendices

Increasingly, plant molecular biologists obtain information from the internet (Harper, 1995). Sometimes the exact details of the information required cannot be prescribed. Fuzzy searching is possible on the worldwide web (Bigwood, 1995).

Some useful internet addresses are given below.

AGRICULTURAL GENOME INFORMATION SERVER

http://probe.nalusda.gov/
Links to animal and plant genome databases.

- Alfagenes – alfalfa (*Medicago sativa*)
- BeanGenes – *Phaseolus* and *Vigna*
- ChlamyDB – *Chlamydomonas reinhardtii*
- CoolGenes – cool season food legumes
- CottonDB – *Gossypium hirsutum*
- GrainGenes – wheat, barley, rye and relatives
- MaizeDB – maize
- PathoGenes – fungal pathogens of small-grain cereals
- RiceGenes – rice
- SolGenes – Solanaceae
- SorghumDB – *Sorghum bicolor*
- SoyBase – soybeans
- TreeGenes – forest trees
- AGRICOLA – plant genetics subset
- CIMMYT – Wheat International Nursery Data
- Ecosys – plant ecological ranges
- EthnobotDB – Native American food plants
- MPNADB – medicinal plants of Native America
- PhytochemDB – plant chemicals
- PVP – Plant Variety Protection

BIODIVERSITY RESOURCE CENTRE

http://www.biodiv.com/biodiv.html
Links to biodiversity resources.

GERMPLASM RESOURCES INFORMATION NETWORK (GRIN)

hppt://www.ars-grin.gov
GRIN facilitates the management and operation of NPGS (National Plant Germplasm System) in the USA. The NPGS preserves the genetic diversity of species important in agriculture and food supply.

GRAINGENES

http://probe.nalusda.gov:8300
Includes information (maps, genes, probes, sequences, pests and specific traits) on wheat, barley, oats, sugarcane and other small grains.

KLOTHO:BIOCHEMICAL COMPOUNDS DECLARATIVE DATABASE

http://ibc.wustl.edu/klotho/
3-D structures of biochemical compounds.

NATIONAL CENTRE FOR BIOTECHNOLOGY INFORMATION (NCBI)

http:///www.ncbi.nlm.nih.gov/
General resources and links for biotechnology.

ARABIDOPSIS DATABASE

http://genome-www.stanford.edu/
General access to data and information on *Arabidopsis*

MENDEL

http://probe.nalusda.gov:8300.html
Analogous genes across the plant kingdom are assigned names corresponding to gene families.

NOTTINGHAM *ARABIDOPSIS* STOCK CENTRE

arabidopsis@nottingham.ac.uk
Information about *Arabidopsis* stocks and maps, how to grow *Arabidopsis*, control of pests and diseases.

PEDRO'S BIOMOLECULAR RESEARCH TOOLS

http://ww.public.iastate.edu/~pedro/research_tools.html
  Molecular biology resources and links.

PHYLOGENY METABOLISM ALIGNMENTS (PUMA)

http:/www.mcs.anl.gov?home/compbio/PUMA/Production?puma_grap
hics.html

PROTEIN DATA BANK

http://www.pdb.bnl.gov/
  3-D structures of macromolecules.

SEQUENCE RETRIEVAL SYSTEM (SRS)

hppt://www.sanger.ac.uk/searching.html
  Browser for molecular databases.

The cloning of genes by homology and the analysis of plant genomes by comparative mapping is aided by an understanding of the relationships between plant groups. The families of flowering plants listed by Cronquist (1988) are given below.

Appendix B: Classification of flowering plants

Outline of Classification of Flowering Plants
(*Magnoliophta*; Cronquist, 1988)

MAGNOLIOPSIDA

Magnoliidae

Magnoliales
Winteraceae
Degeneriaceae
Himantandraceae
Eupomatiaceae
Austrobaileyaceae
Magnoliaceae
Lactoridaceae
Annonaceae

Myristicaceae
Canellaceae

Laurales
Amborellaceae
Trimeniaceae
Monimiaceae (Atherospermataceae, Hortoniaceae, Siparunaceae)
Gomortegaceae
Calycanthaceae
Idiospermaceae
Lauraceae (Cassythaceae)
Hernandiaceae (Gyrocarpaceae)

Piperales
Chloranthaceae
Saururaceae
Piperaceae (Peperomiaceae)

Aristolochiales
Aristolochiaceae

Illiciales
Illiciaceae
Schisandraceae

Nymphaeales
Nelumbonaceae
Nymphaeceae (Euryalaceae)
Barclayaceae
Cabombaceae
Ceratophyllaceae

Ranunculales
Ranunculaceae (Glaucidiaceae, Helleboraceae, Hydrastidaceae)
Circaesteraceae (Kingdoniaceae)
Berberidaceae (Leonticaceae, Nandinaceae, Podophyllaceae)
Sargentodoxaceae
Lardizabalaceae
Menispermaceae
Coriariaceae
Sabiaceae (Meliosmaceae)

Papaverales
Papaveraceae (Chelidoniaceae, Eschscholziaceae, Platystemonaceae)
Fumariaceae (Hypecoaceae, Pteridophyllaceae)

Hamamelidae

Trochodendrales
Tetracentraceae
Trochodendraceae

Hamamelidales
Cercidiphyllaceae
Eupteleaceae
Platanaceae
Hamamelidaceae (Altingiaceae, Rhodoleiaceae)
Myrothamnaceae

Daphniphyllales
Daphniphyllaceae

Didymelales
Didymelaceae

Eucommiales
Eucommiaceae

Urticales
Barbeyaceae
Ulmaceae (Celtidaceae)
Cannabaceae
Moraceae
Cercropiaceae
Urticaceae
Physenaceae

Leitneriales
Leitneriaceae

Juglandales
Rhoipteleaceae
Juglandaceae

Myricales
Myricaceae

Fagales
Balanopaceae
Fagaceae
Nothofagaceae

Betulaceae (Carpinaceae, Corylaceae)

Casuarinales
Casuarinaceae

Caryophyllidae

Caryophyllales (Centrospermae)
Phytolaccaceae (Agdestidaceae, Barbeuiaceae, Gisekiaceae, Petiveriaceae, Stegnospermaceae)
Arhatocarpaceae
Nyctaginaceae
Aizoaceae (Ficoidaceae, Mesembryanthemaceae, Sesuviaceae, Tetragoniaceae)
Didiereaceae
Cactaceae
Chenopodiaceae (Dysphaniaceae, Halophytaceae, Salicorniaceae)
Amaranthaceae
Portulacaceae (Hectorellaceae)
Basellaceae
Molluginaceae
Caryophyllaceae (Alsinaceae, Illecebraceae, Silenaceae)

Polygonales
Polygonaceae

Plumbaginales
Plumbaginaceae (Limoniaceae)

Dilleniidae

Dilleniales
Dilleniaceae
Paeoniaceae

Theales
Ochnaceae (Diegodendraceae, Lophiraceae, Luxemburgiaceae, Strasburgeriaceae, Sauvagesiaceae, Wallaceaceae)
Sphaerosepalaceae (Rhopalocarpaceae)
Sarcolaenaceae
Dipterocaraceae
Caryocarpaceae
Theaceae (Asteropeiaceae, Bonnetiaceae, Camelliaceae, Sladeniaceae, Ternstroemiaceae)
Actinidiaceae (Saurauiaceae)

Scytopetalaceae (Rhaptopetalaceae)
Pentaphylacaceae
Tetrameristaceae
Pellicieraceae
Oncothecaceae
Marcgraviaceae
Quiinaceae
Elatinaceae
Paracryphiaceae
Medusagynaceae
Clusiaceae (Guttiferae*, Garciniaceae, Hypericaceae)

Malvales
Elaeocarpaceae (Aristoteliaceae)
Tiliaceae
Sterculiaceae (Byttneriaceae)
Bombacaceae
Malvaceae

Lecythidales
Lecythidaceae (Asteranthaceae, Barringtoniaceae, Foetidiaceae,
  Napoleonaeaceae)

Nepenthales
Sarraceniaceae
Nepenthaceae
Droseraceae (Dionaeaceae)

Violales
Flacourtiaceae (Berberidopsidaceae, Neumanniaceae,
  Plagiopteridaceae, Soyauxiaceae)
Peridiscaceae
Bixaceae (Cochlospermaceae)
Cistaceae
Huaceae
Lacistemaceae
Scyphostegiaceae
Stachyuraceae
Violaceae (Leoniaceae)
Tamaricaceae
Frankeniaceae
Dioncophyllaceae
Ancistrocladaceae
Turneraceae
Malesherbiaceae

Passifloraceae
Achariaceae
Caricaceae
Fourquieriaceae
Hoplestigmataceae
Cucurbitaceae
Datiscaceae (Tetramelaceae)
Begoniaceae
Loasaceae (Gronoviaceae)

Salicales
Salicaceae

Capparales
Tovariaceae
Capparaceae (Cleomaceae, Koeberliniaceae, Pentadiplandraceae)
Brassicaceae (Cruciferae*)
Moringaceae
Resedaceae

Batales
Gyrostemonaceae
Bataceae

Ericales
Cyrillaceae
Clethraceae
Grubbiaceae
Empetraceae
Epacridaceae (Prionotaceae, Stypheliaceae)
Ericaceae (Vacciniaceae)
Pyrolaceae
Monotropaceae

Diapensiales
Diapensiaceae

Ebenales
Sapotaceae(Achraceae, Boerlagellaceae, Bumeliaceae,
    Sarcospermataceae)
Ebenaceae
Styracaceae
Lissocarpaceae
Symplocaceae

Primulales
Theophrastaceae
Myrsinaceae (Aegicerataceae)
Primulaceae (Coridaceae)

Rosidae

Rosales
Brunelliaceae
Connaraceae
Eucryphiaceae
Cunoniaceae (Baueraceae)
Davidsoniaceae
Dialypetalanthaceae
Pittosporaceae
Byblidaceae (Roridulaceae)
Hydrangeaceae (Kirengeshomaceae, Philadelphaceae, Pottingeriaceae)
Columelliaceae
Grossulariaceae (Argophyllaceae, Brexiaceae, Carpodetaceae,
    Dulongiaceae, Escalloniaceae, Iteaceae, Montiniaceae,
    Phyllonmaceae, Polyosmataceae, Pterostemonaceae, Rousseaceae,
    Tetracarpaeaceae, Tribelaceae)
Greyiaceae
Bruniaceae (Berzeliaceae)
Anisophylleaceae (Polygonanthaceae)
Alseuosmiaceae
Crassulaceae
Cephalotaceae
Saxifragaceae (Eremosynaceae, Francoaceae, Lepuropetalaceae,
    Amygdalaceae, Drupaceae, Malaceae, Pomaceae)
Neuradaceae
Crossosomataceae
Chrysobalanaceae
Surianaceae (Stylobasiaceae)
Rhabdodendraceae

Fabales (Leguminosae*)
Mimosaceae
Caesalpiniaceae
Fabaceae (Papilionaceae*)

Proteales
Elaeagnaceae
Proteaceae

Podostemales
Podostemaceae (Tristichaceae)

Haloragales
Haloragaceae (Myriophyllaceae)
Gunneraceae

Myrtales
Sonneratiaceae (Duabangaceae)
Lythraceae
Rhynchocalycaceae
Alzateaceae
Penaeaceae
Crypteroniaceae
Thymelaeaceae
Trapaceae
Myrtaceae (Heteropyxidaceae, Kaniaceae, Psiloxylaceae)
Punicaceae
Onagraceae
Oliniaceae
Melastomataceae (Memecylaceae, Mouririaceae)
Combretaceae (Strephonemataceae)

Rhizophorales
Rhizophoraceae

Cornales
Alangiaceae
Cornaceae (Aralidiaceae, Aucubaceae, Curtisiaceae, Davidiaceae,
    Griseliniaceae, Helwingiaceae, Mastixiaceae, Melanophyllaceae,
    Nyssaceae, Toricelliaceae)
Garryaceae

Santales
Medusandraceae
Dipentodontaceae
Olacaceae (Aptandraceae, Cathedraceae, Chaunochitonaceae,
    Coulaceae, Erythropalaceae, Heisteriaceae, Octoknemaceae,
    Schoepfiaceae, Scorodocarpaceae, Strombosiaceae, Tetrastylidaceae)
Opiliaceae (Cansjeraceae)
Santalaceae (Anthobolaceae, Canopodaceae, Exocarpaceae,
    Osyridaceae, Podospermaceae)
Misodendraceae
Loranthaceae
Viscaceae

Eremolepidaceae
Balanoporaceae (Cynomoriaceae, Dactylanthaceae, Sarcophytaceae)

Rafflesiales
Hydnoraceae
Mitrastemonaceae
Rafflesiaceae (Apodanthaceae, Cytinaceae)

Celastrales
Geissolomataceae
Celastraceae (Canotiaceae, Chingithamnaceae, Goupiaceae,
    Lophopyxidaceae, Siphonodontaceae)
Hippocrateaceae
Stackhousiaceae
Salvadoraceae
Tepuianthaceae
Aquifoliaceae (Phellinaceae, Sphenostemonaceae)
Icacinaceae (Phtycrenaceae)
Aextoxicaceae
Cardiopteridaceae
Corynocarpaceae
Dichapetalaceae

Euphorbiales
Buxaceae (Pachysandraceae, Stylocerataceae)
Simmondsiaceae
Pandaceae
Euphorbiaceae (Androstachydaceae, Hymenocardiaceae,
    Picrodendraceae, Putranjivaceae, Scepaceae, Stilaginaceae,
    Uapacaceae)

Rhamnales
Rhamnaceae (Camarandraceae, Frangulaceae, Phylicaceae)
Leeaceae
Vitaceae

Linales
Erythroxylaceae (Nectaropetalaceae)
Humiriaceae
Ixonanthaceae
Hugoniaceae (Ctenolophonaceae)
Linaceae

Polygalales
Malpighiaceae

Vochysiaceae
Trigoniaceae
Tremandraceae
Polygalaceae (Diclidantheraceae, Disantheraceae, Emblingiaceae,
   Moutabeaceae)
Xanthophyllaceae
Krameriaceae

Sapindales
Staphyleaceae (Tapisciaceae)
Melianthaceae
Bretschneideraceae
Akaniaceae
Sapindaceae (Ptaeroxylaceae)
Hippocastanaceae
Aceraceae
Burseraceae
Anacardiaceae (Blepharocaryaceae, Pistiaceae, Podoaceae)
Julianiaceae
Simaroubaceae (Irvingiaceae, Kirkiaceae)
Cneoraceae
Meliaceae (Aitoniaceae)
Rutaceae(Flindersiaceae)
Zygophllaceae (Balanitaceae, Nitrariaceae, Peganaceae,
   Tetradiclidaceae, Tribulaceae)

Geraniales
Oxalidaceae (Averrhoaceae, Hypseocharitaceae, Lepidobotryaceae)
Geraniaceae (Biebersteiniaceae, Dirachmaceae, Ledocarpaceae,
   Rhynchothecaceae, Vivianiaceae)
Limnanthaceae
Tropaeolaceae
Balsaminaceae

Apiales
Araliaceae
Apiaceae (Umbelliferae*)

Asteridae

Gentianales
Loganiaceae (Antoniaceae, Desfontainiaceae, Potaliaceae,
   Spigeliaceae, Strychnaceae)
Gentianaceae
Saccifoliaceae

Apocynaceae (Plocospermataceae, Plumeriaceae)
Asclepiadaceae (Periplocaceae)

Solanales
Duckeodendraceae
Nolanaceae
Solanaceae (Goetziaceae, Salpiglossidaceae, Sclerophylacaceae)
Convolvulaceae (Dichondraceae, Humbertiaceae)
Cuscutaceae
Retziaceae
Menyanthaceae
Polemoniaceae (Cobaeaceae)
Hydrophyllaceae

Lamiales
Lennonaceae
Boraginaceae
Verbenaceae (Avicenniaceae, Chloanthaceae, Dicrastylidiaceae,
    Nyctanthaceae, Phrymaceae, Stilbaceae, Symphoremataceae)
    Lamiaceae (Labiatae*, Menthaceae, Tetrachondraceae)

Callitrichales
Hippuridaceae
Callitrichaceae
Hydrostachyaceae

Plantaginales
Plantaginaceae

Scrophulariales
Buddlejaceae
Oleaceae (Fraxinaceae, Syringaceae)
Scrophulariaceae (Ellisiophyllaceae, Rhinanthaceae)
Globulariaceae (Selaginaceae)
Myoporaceae (Spielmanniaceae)
Orobanchaceae
Gesneriaceae (Cyrtandraceae)
Acanthaceae (Thunbergiaceae)
Pedaliaceae (Martyniaceae, Trapellaceae)
Bignoniaceae
Mendonciaceae
Lentibulariaceae (Pinguiculaceae, Utriculariaceae)

Campanulales
Pentaphragmataceae

Sphenocleaceae
Campanulaceae (Cyphiaceae, Cyphocarpaceae, Lobeliaceae,
  Nemacladaceae)
Stylidiaceae
Donatiaceae
Brunoniaceae
Goodeniaceae

Rubiales
Rubiaceae (Henriqueziaceae, Naucleaceae)
Theligonaceae (Cynocramaceae)

Dipsacales
Caprifoliaceae (Carlemanniaceae, Sambucaceae, Viburnaceae)
Adoxaceae
Valerianaceae (Triplostegiaceae)
Dipsacaceae (Morinaceae)

Calycerales
Calyceraceae

Asterales
Asteraceae (Compositae*, Ambrosiaceae, Cichoriaceae)

LILIOPSIDA

Alismatidae

Alismatales
Butomaceae
Limnocharitaceae
Alismataceae

Hydrocharitales
Hydrocharitaceae (Haplophilaceae, Thalassiaceae)

Najadales
Aponogetonaceae
Scheuchzeriaceae
Juncaginaceae (Lilaeceae, Maundiaceae, Triglochinaceae)
Potamogetonaceae
Ruppiaceae
Najadaceae
Zannichelliaceae
Posidoniaceae

Cymodoceaceae
Zosteraceae

Triuridales
Petrosaviaceae
Triuridaceae

Arecidae

Arecales
Arecaceae (Palmae*, Nypaceae, Phytelephasiaceae)

Cyclanthales
Cyclanthaceae

Pandanales
Pandanaceae

Arales
Acoraceae
Araceae
Lemnaceae

Commelinidae

Commelinales
Rapateaceae
Xyridaceae (Abolbodaceae)
Mayacaceae
Commelinaceae (Cartonemataceae)

Eriocaulales
Eriocaulaceae

Restionales
Flagellariaceae
Joinvilleaceae
Restionaceae (Anarthriaceae, Ecdeiocoleaceae)
Centrolepidaceae

Juncales
Juncaceae
Thurniaceae

Cyperales
Cyperaceae (Kobresiaceae)
Poaceae (Gramineae, Anomochloaceae, Bambusaceae,
    Streptochaetaceae)

Hydatellales
Hydatellaceae

Typhales
Sparganiaceae
Typhaceae

Zingiberidae

Bromeliales
Bromeliaceae (Tillandsiaceae)

Zingiberales (Scitamineae)
Strelitziaceae
Heliconiaceae
Musaceae
Lowiaceae (Orchidanthaceae)
Zingiberaceae
Costaceae
Cannaceae
Marantaceae

Liliidae

Liliales
Philydraceae
Pontederiaceae
Haemodoraceae (Conostylidaceae)
Cyanastraceae
Liliaceae (Agapanthaceae, Alliaceae, Alstroemeriaceae,
    Amaryllidaceae, Aphyllanthaceae, Anthericaceae, Asparagaceae,
    Asphodelaceae, Aspidistraceae, Asteliaceae, Blandfordiaceae,
    Calochortaceae, Campynemaceae, Colchicaceae, Convallariaceae,
    Dianellaceae, Eriospermaceae, Funkiaceae, Hemerocallidaceae,
    Herreriaceae, Hesperocallidaceae, Hyacinthaceae, Hypoxidaceae,
    Ixoliriaceae, Medeolaceae, Melanthiaceae, Nartheciaceae,
    Ruscaceae, Tecophilaeaceae, Uvulariaceae)
Iridaceae (Gladiolaceae, Hewardiaceae, Isophysidaceae, Ixiaceae)
Velloziaceae
Aloaceae

Agavaceae (Doryanthaceae, Dracaenaceae, Nolinaceae, Phormiaceae,
  Sansevieriaceae)
Xanthorrhoeaceae (Calectasiaceae, Dasypogonaceae)
Hanguaunaceae
Taccaceae
Stemonaceae (Croomiaceae, Roxburghiaceae)
Smilacaceae (Lapageriaceae, Luzuriagaceae, Petermanniaceae,
  Philesiaceae, Rhipogonaceae)
Dioscoreaceae (Cladophyllaceae, Stenomeridaceae, Tamaceae,
  Trichopodaceae)

Orchidales
Geosiridaceae
Burmanniaceae (Tripterellaceae, Thismiaceae)
Corsiaceae
Orchidaceae (Apostasiaceae, Cypripediaceae, Limodoraceae,
  Neottiaceae, Thyridiaceae, Vanillaceae)

*Alternative names

List of species accounting for a significant proportion of the weight,
calories, protein or fat in human diets on a national basis (Prescott-
Allen and Prescott-Allen, 1995). The molecular biology of these species
is likely to find application in the improvement of world agriculture
and food supply.

> **Appendix C:**
> **Plants representing**
> **important sources**
> **of food**

Commodity	Species
Wheats	*Triticum aestivum* (L.) Thell., *T. turgidum* (L.) Thell
Rices	*Oryza glaberrima* Steud, O. *sativa* L.
Maize	*Zea mays* L.
Sorghum	*Sorghum bicolor* (L.) Moench
Millets	*Echinochloa frumentacea* Link, *Eleusine cora-cana* (L.) Gaertner, *Panicum miliaceum* L., *Pennisetum americanum* (L.) Leeke, *Setaria italica* (L.) Pal.
Rye	*Secale cereale* L.
Barley	*Hordeum vulgare* L.
Oats	*Avena sativa* L.
Fonio	*Digitaria exilis* Stapf
Quinoa	*Chenopodium quinoa* Willd
Potato	*Solanum tuberosum* L.
Cassava	*Manihot esculenta* Crantz
Yams	*Dioscorea alata* L., *D. cayenensis* Lam./D.

	rotundata Poiret, *D. dumetorum* (Kunth) Pax, *D. esculenta* (Lour.) Burkill, *D. trifida* L.f.
Sweet potato	*Ipomoea batatas* (L.) Lam.
Taro	cocoyam, *Colocasia esculenta* (L.) Schott
Yautia	*Xanthosoma sagittifolium* (L.) Schott
Sugarcane	*Saccharum officinarum* L. & sugar beet *Beta vulgaris* L.
Soybean	*Glucine max* (L.) Merr.
Groundnut	*Arachis hypogaea* L.
Beans (dry and green)	*Lablab purpureus* (L.) Sweet,
Beans (sweet)	*Phaseolus lunatus* L., *P. vulgaris* L., *Vigna angularis* (Willd.) Ohwi & Ohashi, *V. radiata* (L.) R. Wilczek.
Cowpea (dry)	*Vigna unguiculata* (L.) Walp.
Pea (dry and green)	*Pisum sativum* L.
Pigeonpea	*Cajanus cajan* L. Millsp.
Chickpea	*Cicer arietinum* L.
Broad bean	*Vicia faba* L.
Lentil	*Lens culinaris* Medik
Lupin	*Lupinus mutabilis* Sweet
Coconut	*Cocos nucifera* L.
Sunflower seed	*Helianthus annuus* L.
Oil palms	*Elaeis guineensis* Jacq.
Cottonseeds	*Gossypium barbadense* L., *G. hirsutum* L.
Olive	*Olea europaea* L.
Rapeseeds	*Brassica napus* L., *B. rapa* L.
Sesame seeds	*Sesamum orientale* L.
Melon seeds	*Citrullus lanatus* (Thunb) Matsum. & Nakai, *Cucumis melo* L.
Karite nut, sheanut	*Vitellaria paradoxa* Gaertner f.
Almond	*Prunus dulcis* (Miller) D. Webb
Filbert	*Corylus avellana* L./*C. maxima* Miller
Mustard seed	*Brassica juncea* (L.) Czerniak
Safflower seed	*Carthamus tinctorius* L.
Walnut	*Juglans regia* L.
Brazil nut	*Bertholletia excelsa* Bonpl.
Pistachio	*Pistacia vera* L.
Tomato	*Lycopersicon esculentum* Miller
Cabbages	*Brassica oleracea* L.B. *rapa* L.
Onions (green and dry)	*Allium cepa* L., *A. fistulosum* L.
Carrot	*Daucus carota* L.
Cucumber	*Cucumis sativus* L.
Pumpkins, squash, gourds	*Cucurbita maxima* Duchesne, *C. moschata* (Duchesne) Poiret, *C. pepo* L.

Lettuce	*Lactuca sativa* L.
Chilli pepper, sweet pepper	*Capsicum annuum* L.
Eggplant	*Solanum melongena* L.
Garlic	*Allium sativum* L.
Spinach	*Spinacia oleracea* L.
Artichoke	*Cynara scolymus* L.
Banana and plantain	*Musa acuminata* Colla/M. × *paradisiaca* L.
Apple	*Malus pumila* Miller
Orange	*Citrus sinensis* (L.) Osbeck
Grape	*Vitis vinifera* L.
Watermelon	*Citrullus lanatus* (Thunb.) Matsum. & Nakai
Date	*Phoenix dactulifera* L.
Avocado	*Persea americana* Miller
Mango	*Mangifera indica* L.
Pineapple	*Ananas comosus* (L.) Merr
Tangerine, mandarin	*Citrus reticulata* Blanco
Lemon	*Citrus limon* (L.) Burm.f & lime, *C auranti-ifolia* (Christm.) Swingle
Grapefruit	*Citrus* × *paradisi* Macfad./pomelo *C. grandis* (L.) Osbeck
Melon	*Cucumis melo* L.
Papaya	*Carica papaya* L.
Pear	*Pyrus communis* L.
Peach, nectarine	*Prunus persica* (L.) Batsch
Plum	*Prunus domestica* L.
Fig	*Ficus carica* L.
Strawberry	*Fragaria* × *ananassa* Duchesne
Apricot	*Prunus armeniaca* L.
Cherry	*Prunus avium* (L.) L.
Currants	*Ribes nigrum* L. or *R rubrum* L.
Pimento, allspice	*Pimenta dioica* (L.) Merr. or *Capsicum annuum* L.
Star anise	*Illicium verum* Hook. f.
Cardamom	*Elettaria cardamomum* (L.) Maton
Pepper	*Piper nigrum* L.
Cocoa	*Theobroma cacao* L.
Coffees	*Coffea arabica* L., *C. canephora* Pierre
Mate	*Ilex paraguariensis* A. St-Hil.
Tea	*Camellia sinensis* (L.) Kuntze

Appendix D: Nuclear DNA content of important plant species

(As determined by flow cytometry; Arumuganathan and Earle, 1991.)
    More detail is available on the internet at the USDA, ARS National Agricultural Library site.

Species		Nuclear DNA content
Common name	Scientific name	~Mbp/1C
ND	*Aegilops squarrosa*	4024
Leek	*Allium ampeloprasum*	24255
Onion	*Allium cepa*	15 290–15 797
Red pineapple	*Ananas bracteatus*	444
Pineapple	*Ananas comosus*	526
Arabidopsis	*Arabidopsis thaliana*	145
Peanut/groundnut	*Arachis hypogaea* (2n = 4X)	2813
Asparagus	*Asparagus officinalis*	1308
Oats	*Avena sativa*	11315
Beet/beetroot	*Beta vulgaris* ssp. *esculenta*	714
Sugar beet	*Beta vulgaris* ssp. *saccharifera*	758
Pak choi	*Brassica campestris* ssp. *chinensis*	507
Turnip rape	*Brassica campestris* ssp. *oleifera*	468–516
Turnip	*Brassica campestris* ssp. *rapifera*	511
White mustard	*Brassica hirta* (= *Sinapis alba*)	492
Brown mustard	*Brassica juncea*	1105
Rapeseed	*Brassica napus*	1129–1235
Black mustard	*Brassica nigra*	468
Cauliflower	*Brassica oleracea* ssp. *botrytis*	628-662
Cabbage	*Brassica oleracea* ssp. *capitata*	603
Brussels sprouts	*Brassica oleracea* ssp. *gemmifera*	628
Broccoli	*Brassica oleracea* ssp. *italica*	599–618
ND	*Brassica tourneforthii*	791
Pepper/chilli	*Capsicum annuum*	2702–3420
Papaya	*Carica papaya*	372
Chick pea	*Cicer arietinum*	738
Watermelon	*Citrullus vulgaris* (= *lanatus*)	425, 434
Orange	*Citrus sinensis*	367, 396
Crepis	*Crepis capillaris*	1867
Cantaloupe	*Cucumis melo*	454, 502
Cucumber	*Cucumis sativus*	367
Zucchini	*Cucurbita pepo*	502, 521
Jimson weed	*Datura stramoniun*	1983
Carrot	*Daucus carota*	473
Yam	*Dioscorea alata*	555
ND	*Diplotaxis erucoides*	632

ND	*Eruca sativa*	560
Soybean	*Glycine max* (2n = 4X)	1115
Cotton	*Gossypium hirsutum* (2n = 4X)	2118–2374
Sunflower	*Helianthus annuus*	2871–3189
Barley	*Hordeum vulgare*	4873
Sweet potato	*Ipomoea batatas* (2n = 6X)	1597
Lettuce	*Lactuca sativa*	2639
Lentil	*Lensculinaris (= esculenta)*	4063
ND	*Lycopersicon cheesemanii*	883
Tomato	*Lycopersicon esculentum*	907–1000
ND	*Lycopersicon pennellii*	1192–1337
ND	*Lycopersicon peruvianum*	1095
Apple	*Malus x domestica* (2n = 2X)	743–796
Mango	*Mangifera indica*	439
Cassava/manioc	*Manihot esculenta (= utilissima)*	690–830
Alfalfa/lucerne	*Medicago sativa* (2n = 4X)	1510
ND	*Medicago truncatula*	454–526
Sweet clover	*Melilotus officinalis*	1086
Banana	*Musa* sp.	873
ND	*Nicotiana plumbaginifolia*	2287
Tobacco	*Nicotiana tabacum* (2n = 4X)	4221–4646
African rice	*Oryza longistaminata*	376
Rice	*Oryza sativa* ssp. *Indica*	419–463
Rice	*Oryza sativa* ssp. *Japonica*	415–439
Rice	*Oryza sativa* ssp. *Javanica*	424
Passion fruit	*Passiflora menspermifolia*	2191
Avocado	*Persea americana*	883
Parsley	*Petroselinum crispum*	1911
Petunia	*Petunia hybrida*	1274
ND	*Petunia parodii*	1221
Tepary bean	*Phaseolus acutifolius*	647
Scarlet bean	*Phaseolus coccineus runner*	709
Lima bean	*Phaseolus lunatus*	622
Common bean	*Phaseolus vulgaris*	637
Garden pea	*Pisum sativum*	3947–4397
Apricot	*Prunus armenaica*	294
Sweet cherry	*Prunus avium*	338
Cherry	*Prunus avium × cerasus* (2n = 4X)	685
Sour cherry	*Prunus cerasus* (2n = 4X)	599
Prune	*Prunus domestica* (2n = 6X)	883
Peach	*Prunus persica*	262–265
Pear	*Pyrus communis*	496–536
Radish	*Raphanus sativus*	526
Castor bean	*Ricinus communis*	323

Raspberry	*Rubus idaeus*	280
Sugarcane	*S. barberi × S. spontaneum*	2953
Sugarcane	*Saccharum barberi*	3156–4121
Sugarcane	*Saccharum officinarum*	2547–3605
Sugarcane	*Saccharum robustum*	3151
Sugarcane	*Saccharum sinense*	4183
ND	*Sesbania rostrata*	1187
ND	*Sinapis arvensis*	367
ND	*Solanum berthaultii*	840
Eggplant	*Solanum melongena*	1100, 1197
Potato	*Solanum tuberosum* (2n = 4X)	1597–1862
Sorghum	*Sorghum bicolor*	748–772
Spinach	*Spinacia oleracea*	989
Red clover	*Trifolium pratense*	468
White clover	*Trifolium repens*	999
Gama grass	*Tripsacum dactyloides*	3730
Wheat	*Triticum aestivum* (2n = 6X)	15 966
ND	*Triticum monococcum*	5751
Garden tulip	*Tulipa* sp.	24 704–30 687
Vanilla	*Vanilla planifolia*	7672
Black gram/urud	*Vigna mungo*	574
Mung bean	*Vigna radiata*	579
Cowpea	*Vigna unguiculata* (= *sinensis*)	613
Grape	*Vitis vinifera*	483
Calamondin orange	*X Citrofortunella mitis*	386
ND	*Zea diploperennis*	1723
Corn	*Zea mays*	2292–2716

**Appendix E:
Patentable subject
matter in plant
molecular biology**

The following list includes some of the subjects that may be the subject of patents. Application of plant molecular biology requires a knowledge of such patents.

- DNA isolated from any source
- Proteins resulting from the expression of isolated DNA
- Diagnostic probes
- Genetic traits or markers
- Transgenic plants

An invention can only be patented if it is not obvious (Finney, 1994).

Patented process are sometimes licensed with the sale of reagents or kits. This is especially common in relation to research use of patented molecular biology procedures. Commercial exploitation usually requires separate specific licensing.

Plant genotypes may also be protected by Plant Breeders' Rights in some countries. This is a specific type of plant patent designed to reward innovation in plant breeding. The propagation of protected genotypes requires approval of the holder of the rights to the genotype.

The following approach to recording of research may help support applications for patent protection and is good laboratory practice.

**Appendix F:
Guidelines for
laboratory
notebooks**

- Always use bound notebooks with consecutively numbered pages. Draw a line through unused portions of pages. Do not leave blank pages.
- Always write in permanent ink. Do not use pencils. Do not erase or use correcting fluid. Explain all corrections in the book.
- Entries should include the following:
  The date on which the experiment was started;
  Why the experiment was done;
  A description of the methods and reagents used;
  Identification of the individuals involved in the work;
  The results obtained; and
  Conclusions drawn from the results.
- Always record negative results (not only positive ones).
- Sign and date each page daily or as the page is completed.
- Each page should be countersigned. It is important that the person countersigning is not a co-inventor. The person countersigning must understand the technical details of the material in the notebook, but is not attesting to the accuracy of the result in the notebook and need not have seen the experiment conducted.

Some useful reference material for plant molecular biology is given below.

**Appendix G:
Reference
information for
plant molecular
biology**

CODES FOR REDUNDANT NUCLEOTIDES IN SEQUENCES

Pu   R = A,G
Py   Y = C,T
      M = A,C
      K = G,T
      S = G,C
      W = A,T

      H = A,T,C
      B = G,T,C
      V = G,A,C
      D = G,A,T

      N = G,A,T,C

CODONS

	Second position			
	T	C	A	G
T	TTT Phe	TCT Ser	TAT Tyr	TGT Cys
First position	TTC Ser	TCC Ser	TAC Tyr	TGC Cys
	TTA Leu	TCA Ser	TAA Stop	TGA Stop
	TTG Leu	TCG Ser	TAG Stop	TGG Trp
C	CTT Leu	CCT Pro	CAT His	CGT Arg
	CTC Leu	CCC Pro	CAC His	CGC Arg
	CTA Leu	CCA Pro	CAA Gln	CGA Arg
	CTG Leu	CCG Pro	CAG Gln	CGG Arg
A	ATT Ile	ACT Thr	AAT Asn	AGT Ser
	ATC Ile	ACC Thr	AAC Asn	AGC Ser
	ATA Ile	ACA Thr	AAA Lys	AGA Arg
	ATG Met	ACG Thr	AAG Lys	AGG Arg
G	GTT Val	GCT Ala	GAT Asp	GGT Gly
	GTC Val	GCC Ala	GAC Asp	GGC Gly
	GTA Val	GCA Ala	GAA Glu	GGA Gly
	GTG Val	GCG Ala	GAG Glu	GGG Gly

AMINO ACIDS

Alanine	Ala	A
Arginine	Arg	R
Asparagine	Asn	N
Aspartic acid	Asp	D
Cysteine	Cys	C
Glutamine	Gln	Q
Glutamic acid	Glu	E
Glycine	Gly	G
Histidine	His	H
Isoleucine	Ile	I
Leucine	Leu	L
Lysine	Lys	K
Methionine	Met	M
Phenylalanine	Phe	F
Proline	Pro	P
Serine	Ser	S
Threonine	Thr	T
Tryptophan	Trp	T
Tyrosine	Tyr	Y

Valine   Val   V
Hydroxylysine   Hyl
Hydroxyproline   Hyp

PREFIXES FOR UNITS

10^{-18} atto   a
10^{-15} femto   f
10^{-12} pico   p
10^{-9} nano   n
10^{-6} micro   μ
10^{-3} milli   m
10^{-2} centi   c
10^{-1} deci   d
10   deka   da
10^{2} hecto   h
10^{3} kilo   k
10^{6} mega   m
10^{9} giga   g
10^{12} tera   t

COMMON ABBREVIATIONS FOR NUCLEIC ACIDS

deoxyribonucleic acid   DNA
complementary DNA   cDNA
chloroplast DNA   ctDNA
mitochondrial DNA   mtDNA
nuclear DNA   nDNA
plastid DNA   ptDNA
ribonucleic acid   RNA
messenger RNA   mRNA
ribosomal RNA   rRNA
transfer RNA   tRNA

INTERNATIONAL SOCIETY FOR PLANT MOLECULAR BIOLOGY

The International Society for Plant Molecular Biology (ISPMP) orga-
nizes regular international congresses. ISPMB may be contacted at
the Department of Biochemistry, University of Georgia 30602 USA;
Telephone 1-706-542-2086 or Ldure@uga.cc.uga.edu

PLANT GENOME CONFERENCES

A series of meetings on the plant genome have been held each year since 1993 in San Diego California. These meetings provide an opportunity to review progress in the study of plant genomes.

RESTRICTION ENZYMES

Restriction endonucleases are important tools for the cloning of plant genes and for the analysis of plant genomes. Under some conditions the specificity of some enzymes may be reduced resulting in 'star activity'. Methylation inhibits the action of some enzymes. Most *E. coli* host strains encode DNA methylating systems: *dam* (methylating adenine) or *dcm* (methylating cytosine) in specific sequences. Commercial suppliers such as New England BioLabs and Boehringer-Mannheim produce catalogues that are good sources of information on restriction enzymes.

Some restriction endonucleases

Enzyme	Source	Recognition sequence
*Aat*II	*Acetobacter aceti*	5' GACGT/C 3' 3' C/TGCAG 5'
*Aci*I	*Arthrobacter citreus*	5' C/CGC 3' 3' GGC/G 5'
*Afl*II	*Anabaena flosaquae*	5' C/TTAAG 3' 3' GAATT/C 5'
*Age*I	*Agrobacterium gelatinovorum*	5' A/CCGGT 3' 3' TGGCC/A 5'
*Alu*I	*Arthrobacter luteus*	5' AG/CT 3' 3' TC/GA 5'
*Alw*I	*Acinetobacter lwoffi*	5' GGATC(N_4) 3' 3' CCTAG(N_5) 5'
*Apa*I (*Bsp*120 I)	*Acetobacter pasteurianus*	5' GGGCC/C 3' 3' C/CCGGG 5'
*Ase*I	*Aquaspirillum serpens*	5'AT/TAAT 3' 3' TAAT/TA 5'
*Bam*HI	*Bacillus amyloquefaciens* H	5' G/GATCC 3' 3' CCTAG/G 5'

*Bgl*I	*Bacillus globigii*	5' GCCNNNN/NGGC 3' 3' CGGN/NNNNCCG 5'
*Bgl*II	*Bacillus globigii*	5' A/GATCT 3' 3' TCTAG/A 5'
*Bsp*143 I (*Sau*3AI)	*Bacillus* species RFL143	5'/GATC 3' 3' CTAG/5'
*Bsu*15 I (*Cla*I)	*Bacillus subtilis*	5' AT/CGAT 3' 3' TA GC/TA 5'
*Dra*I	*Deinococcus radiophilus*	5' TTT/AAA 3' 3' AAA/TTT 5'
*Eco*47 I (*Ava*II)	*Escherichia coli* 47	T 5'G/ G ACC3' 3' C C TG/G5' A
*Eco*88 I (*Ava*I)	*Escherichia coli* 88	5' C/PyCGPvG 3' 3' Gpu GCPy/C 5'
*Eco*91 I (*Bst*EII)	*Escherichia coli* 91	5' G/GTNACC 3' 3' CCANTG/G 5'
*Eco*105 I (*Sna*BI)	*Escherichia coli* 105	5' TA C/GTA 3' 3' AT G/CAT 5'
*Eco*130 I (*Sty*I)	*Escherichia coli* 130	AA 5' C/C TTGG 3' 3' G G AA/CC 5' TT
*Eco*RI	*Escherichia coli*	5' G/AATTC 3' 3' CTTAA/G 5'
*Eco*RV (*Eco*32 I)	*Escherichia coli*	5' GAT/ATC 3' 3' CTA/TAG 5'
*Ehe*I (*Nar*I)	*Erwinia herbicola*	5' GGC/GCC 3' 3' CCG/CGG 5'
*Hae*II	*Haemophilus aegyptius*	5' PuGCGC/Py 3' 3' Py/CGCGPu 5'
*Hae*III (*Bsu*R I)	*Haemophilus aegyptius*	5' GC/CC 3' 3'CC/GG 5'
*Hha*I (*Hin*6I)	*Haemophilus haemolyticus*	5' GCG/C 3' 3' C/GCG 5'

*Hinc*II	*Haemophilus influenzae*	5' GTPy/PuAC 3' 3' CAPu/PyTG 5'
*Hind*III	*Haemophilus influenzae*	5' A/AGCTT 3' 3' TTCGA/A 5'
Hinf I	*Haemophilus influenzae*	5' G/ANTC 3' 3' CTNA/G 5'
*Hpa*I	*Haemophilus parainfluenzae*	5' GTT/AAC 3' 3' CAA/TTG 5'
*Hpa*II	*Haemophilus parainfluenzae*	5' C/CGG 3' 3' GGC/C 5'
*Kpn*I	*Klebsiella pneumonia*	5' GGTAC/C 3' 3' C/CATGG 5'
*Mbo*I (*Nde*II)	*Morexella bovis*	5' /GATC 3' 3' CTAG/ 5'
*Mlu*I	*Micrococcus luteus*	5' A/CGCGT 3' 3' TGCGC/A 5'
*Msp*I	*Moraxella species*	5' C/CGG 3' 3' GGC/C 5'
*Mva*I (*Bst*NI)	*Micrococcus varians*	T 5' CC/ AGG 3' 3' GG T/CC 5' A
*Nco*I	*Nocardia corallina*	5' C/CATGG 3' 3' GGTAC/C 5'
*Nde*I	*Neisseria donitrificans*	5' CA/TATG 3' 3' GTAT/AC 5'
*Nla*IV	*Neisseria lactamia*	5' GGN/NCC 3' 3' CCN/NGG 5'
*Not*I	*Nocardia otitidiscavarum*	5' TCG/CGA 3' 3' AGC/GCT 5'
*Nru*I	*Nocardia rubra*	5' TCG/CGA 3' 3' AGC/GCT 5'
*Nsi*I	*Neisseria sicca*	5' ATGCA/T 3' 3' T/ACGTA 5'
*Pac*I	*Pseudomonas alcaligenes*	5' TTAAT/TAA 3' 3' AAT/TAATT 5'

*Pae*I (*Sph* I)	*Pseudomonas aeruginosa*	5' GCATG/C 3' 3' C/GTAC/G 5'
*Ple*I	*Pseudomonas lemoignei*	5' GAGTC(N₄)/ 3' 3' CTCAG(N₅)/ 5'
*Pme*I	*Pseudomonas medicina*	5' GTTT/AAAC 3' 3' CAAA/TTTG 5'
*Pml*I (*Eco*72I)	*Pseudomonas maltophila*	5' CAC/GTG 3' 3' GTG/CAC 5'
*Pst*I	*Providencia stuartii*	5' CTGCA/G 3' 3' G/ACGTC 5'
*Pvu*I	*Proteus vulgaris*	5' CGAT/CG 3' 3' GC/TAGC 5'
*Pvu*II	*Proteus vulgaris*	5' CAG/CTG 3' 3' GTC/GAC 5'
*Rsa*I (*Ccp*6I)	*Rhodopseudomonas sphaeroides*	5' GT/AC 3' 3' CA/TG 5'
*Sac*I (*Ecl*136I)	*Streptomyces achromogenes*	5' GAGCT/C 3' 3' C/TCGAG 5'
*Sac*II (*Cfr*42I)	*Streptomyces achromogenes*	5' CCGC/GG 3' 3' GG/CGCC 5'
*Sal*I	*Streptomyces albus*	5' G/TCGAC 3' 3' CAGCT/G 5'
*Sca*I	*Streptomyces caespitosus*	5' AGT/ACT 3' 3' TCA/TGA 5'
*Sma*I	*Serratia marcescens*	5' CCC/GGG 3' 3' GGG/CCC 5'
*Ssp*I	*Sphaerotilus species*	5' AAT/ATT 3' 3' TTA/TAA 5'
*Stu*I (*Eco*147I)	*Streptomyces tubercidicus*	5' AGG/CCT 3' 3' TCC/GGA 5'
*Taq*I	*Thermus aquaticus*	5' T/CGA 3' 3' AGC/T 5'
*Xba*I	*Xanthomonas badrii*	5' T/CTAGA 3' 3' AGATC/T 5'
*Xho*I	*Xanthhomonas holcicola*	5' C/TCGAG 3' 3' GAGCT/C 5'

*Xma*I   *Xanthomonas malvacerum*   5' C/CCGGG 3'
                                          3' GGGCC/C 5'

Isoschizomers are different enzymes that recognize the same sequence. However, the cleavage point within the sequence may not always be the same, resulting in differing terminal sequences.

BUFFERS

Tris [tris (hydroxymethyl) aminomethane] is the most widely used buffer in plant molecular biology.

Tris has a pK_a of 8.85 at 0°C, 8.06 at 25°C and 7.72 at 37°C and thus the pH of a Tris buffer can fall by more than 1 pH unit if a buffer is warmed from 0°C to 37°C (Perrin *et al.*, 1974)

NOMENCLATURE OF PLANT GENES

The International Society for Plant Molecular Biology (ISPMB) Commission on Plant Gene Nomenclature provides standard names for plant genes that can be applied to all species. A collation of reports of the Commission can be found in *Plant Molecular Biology Reporter*, Volume 12, Number 2, Supplement 1994. The system is based upon the association between genes encoding similar products throughout the plant kingdom. The system classifies gene families. Examples are as follows:

Starch-degrading enzymes (Smith-White and Preiss, 1994)

Gene product	Mnemonic	Product number
α-amylase	*Amy1*	2.3.2.1.1.1
	Amy2	2.3.2.1.1.2
	Amy3	2.3.2.1.1.3
β-amylase	*Bmy1*	2.3.2.1.2

Proteinase inhibitors (Xavier-Filho and Paiva Campos, 1994)

Class	Family	Mnemonic	Product number
Serine	Bowman–Birk inhibitor	*Pis1*	4.4.1.1.1.1
	soybean trypsin inhibitor	*Pis2*	4.4.1.1.1.2
	potato inhibitor I	*Pis3*	4.4.1.1.1.3
	potato inhibitor II	*Pis4*	4.4.1.1.1.4
	squash trypsin inhibitor	*Pis5*	4.4.1.1.1.5
	cereal superfamily	*Pis6*	4.4.1.1.1.6

ragi/maize bifunctional inhibitor		*Pis7*	4.4.1.1.1.7
thaunatin-PR-like		*Pis8*	4.4.1.1.1.8
Cysteine	phytocystatin	*Pic1*	4.4.1.1.2.1
Metallo-	potato carboxypeptidase inhibitor	*Pim1*	4 . 4 . 1 . 1 . 3 . 1
Aspartic	potato aspartyl proteinase	*Pis1*	4.4.1.1.4.1

Mnemonics are composed of three letters followed by a number.

Gene product numbers include the EC number for the enzyme (International Union of Biochemistry) for gene products with enzyme activity.

PLANT CULTURE MEDIA (Data from Mantell *et al.*, 1985)

Component concentration	MS	B5	White	Heller
(mg/l)				
$(NH_4)_2SO_4$	–	134	–	–
$(NH_4)NO_3$	1650	–	–	–
$NaNO_3$	–	–	–	600
KNO_3	1900	2500	80	–
$Ca(NO_3)_2$	–	–	300	–
$CaCl_2.2H_2O$	440	150	–	75
$MgSO_4.7H_2O$	370	250	720	250
Na_2SO_4	–	–	200	–
KH_2PO_4	170	–	–	125
$NaH_2PO_4.H_2O$	–	150	16.5	–
KCl	–	–	65	750
$FeSO_4.7H_2O$	27.8	27.8	–	–
Na_2EDTA	37.3	37.3	–	–
$FeCl_3.6H_2O$	–	–	–	1.0
$Fe_2(SO_4)_3$	–	–	2.5	–
$MnSO_4.4H_2O$	22.3	–	7	0.01
$MnSO_4.H_2O$	–	10	–	–
$ZnSO_4.7H_2O$	8.6	2	3	1
H_3BO_4	6.2	3	1.5	1
KI	0.83	0.75	0.75	0.01
$Na_2MoO_4.2H_2O$	0.25	0.25	–	–
$CuSO_4.5H_2O$	0.025	0.025	–	0.03
$CoCl_2.6H_2O$	0.025	0.025	–	–

NiCl$_2$.6H$_2$O	–	–	–	0.03
AlCl$_3$	–	–	–	0.03
Myo-inositol	100	100	–	–
Nicotinic acid	0.5	1.0	0.5	–
Pyridoxine HCl	0.5	1.0	0.1	–
Thiamine HCl	0.1	10.0	0.1	1.0
Glycine	2.0	–	3.0	–
Ca D-pantothenic acid	–	–	1.0	–
Sucrose	30 000	20 000	20 000	20 000
Kinetin	0.04–10	0.1	–	–
2,4-D	–	0.1–1.0	6.0	–
Indoleacetic acid (IAA)	1.0–30	–	–	–
pH (before autoclaving)	5.7–5.8	5.5	5.5	–

References

Arumuganathan, K. and Earle, E.D. (1991) Nuclear DNA content of some important plant species. *Plant Molecular Biology Reporter*, 9, 211–15.

Bigwood, D.W. (1995) Using fuzzy searching to retrieve plant genome information at the USDA, national agricultural library. *Plant Molecular Biology Reporter*, 13, 6–17.

Cronquist, A. (1988) *The Evolution and Classification of Flowering Plants*, Second Edition, The New York Botanical Garden, New York, pp. 503–17.

Finney, K. (1994) Inventions in biotechnology and the assessment of obviousness. *Australasian Biotechnology*, 4, 280–3.

Harper, R. (1995) World wide web resources for the biologist. *Trends in Genetics*, 11, 223–8.

Mantell, S.H., Matthews, J.A. and McKee, R.A. (1985) *Principles of Plant Biotechnology. An Introduction to the Genetic Engineering of Plants*. Blackwell Scientific Publications, Oxford.

Perrin, D.D. and Dempsey, B. (1974) *Buffers for pH and Metal Ion Control*. Chapman & Hall, London.

Prescott-Allen, R. and Prescott-Allen, C. (1995) How many plants feed the world? in *Readings from Conservation Biology* (ed. D. Ehrenfeld), Blackwell Science Inc., pp. 72–81.

Smith-White, B. and Preiss, J. (1994) Suggested mnemonics for cloned DNA corresponding to enzymes involved in starch metabolism. *Plant Molecular Biology Reporter*, 12, S67–71

Xavier-Filho, J. and Paiva Campos, F.A. (1994) Genes encoding protease inhibitors. *Plant Molecular Biology Reporter*, 12, S58–9.

Index